Principles of Biophotonics, Volume 2

Light emission, detection, and statistics

IPEM–IOP Series in Physics and Engineering in Medicine and Biology

About the Series

Series in Physics and Engineering in Medicine and Biology will allow IPEM to enhance its mission to 'advance physics and engineering applied to medicine and biology for the public good.'

Focusing on key areas including, but not limited to:

- clinical engineering
- diagnostic radiology
- informatics and computing
- magnetic resonance imaging
- nuclear medicine
- physiological measurement
- radiation protection
- radiotherapy
- rehabilitation engineering
- ultrasound and non-ionising radiation.

A number of IPEM–IOP titles are published as part of the EUTEMPE Network Series for Medical Physics Experts.

Principles of Biophotonics, Volume 2

Light emission, detection, and statistics

Gabriel Popescu

Department of Electrical and Computer Engineering, Beckman Institute for Advanced Science and Technology, University of Illinois at Urbana-Champaign, Illinois, USA

IOP Publishing, Bristol, UK

ISBN 978-0-7503-1644-6 (ebook)
ISBN 978-0-7503-1642-2 (print)
ISBN 978-0-7503-1951-5 (myPrint)
ISBN 978-0-7503-1643-9 (mobi)

DOI 10.1088/978-0-7503-1644-6

Version: 20191101

IOP ebooks

British Library Cataloguing-in-Publication Data: A catalogue record for this book is available from the British Library.

Published by IOP Publishing, wholly owned by The Institute of Physics, London

IOP Publishing, Temple Circus, Temple Way, Bristol, BS1 6HG, UK

US Office: IOP Publishing, Inc., 190 North Independence Mall West, Suite 601, Philadelphia, PA 19106, USA

To Catherine, Sophia, Sorin, and my mother.

Motto

'An experiment is a question which science poses to Nature, and a measurement is the recording of Nature's answer.'

Max Planck

Contents

Preface to Volume 2: 'Light emission, detection, and statistics'

This volume describes the most common properties of light, mechanisms of light emission, various approaches for optical detection, and the basic introduction to field correlations and statistical optics.

Chapter 1 places the optical spectrum in the broad context of electromagnetic radiation. The radiometric properties of light are introduced in chapter 2, their respective photon-based quantities are described in chapter 3, and the photometric quantities in chapter 4. Many of these properties will be used later when describing biophotonics methods and their capabilities. Light emission is presented in chapter 5 (fluorescence), chapter 6 (black body radiation) and chapter 7 (laser). The fundamental differences between these light sources will be encountered many times throughout the book. Chapters 8–14 are dedicated to optical detection, covering the various types of detectors, their principles of operation and figures of merit. Choosing the right detector for a particular instrument is crucial in biophotonics, as we will see later in subsequent volumes. Finally, chapter 15 introduces the basic concepts of statistical optics, which we will also encounter later, particularly when discussing investigation modalities based on light interference. As we will find out, an image is itself just a complicated interferogram. Thus, the concepts described in chapter 15 will also help us understand various imaging modalities.

Each chapter benefits from a set of problems, which invite the reader to review and test the important concepts. These problem sets will hopefully help the teacher in using the book in the classroom. Finally, the references provide avenues for more in-depth study on the topic of each chapter.

Once they have mastered the mathematical tools in Volume 1 and the basic concepts of emission, detection, and statistics in Volume 2, the reader will be well prepared to follow the later volumes on light propagation in various types of materials.

Gabriel Popescu
Urbana, Illinois
August 2019

Acknowledgments

I am grateful to my teachers, colleagues, and students, from whom I have been learning every day. In preparation of the manuscript, I received generous support from the ECE Department at UIUC.

I would like to acknowledge help with typesetting and figures from Ionut Preoteasa, and with proof reading by Chenfei Hu and Michael Fanous. Finally, I am grateful to the IOP for their assistance, especially to Jessica Fricchione, Sarah Armstrong, and the Production team.

Author biography

Gabriel Popescu

Gabriel Popescu is a Professor in Electrical and Computer Engineering, University of Illinois at Urbana-Champaign. He received his PhD in Optics in 2002 from the School of Optics/CREOL (now the College of Optics and Photonics), University of Central Florida and continued his training with Michael Feld at M.I.T., working as a postdoctoral associate. He joined Illinois in August 2007, where he directs the Quantitative Light Imaging Laboratory (QLI Lab) at the Beckman Institute for Advanced Science and Technology. Dr Popescu served as Associate Editor of the journals *Optics Express* and *Biomedical Optics Express*, and as an Editorial Board Member for *Journal of Biomedical Optics and Scientific Reports*. He is an OSA and SPIE Fellow.

IOP Publishing

Principles of Biophotonics, Volume 2
Light emission, detection, and statistics

Gabriel Popescu

Chapter 1

Electromagnetic fields

1.1 Regions of the electromagnetic spectrum

Biophotonics employs electromagnetic fields to study biological specimens. While numerous biophotonics tools operate with *visible* light, for example, the light microscope, others use electromagnetic fields outside the visible range. We will study thoroughly in the subsequent volumes electromagnetic field propagation and their interaction with matter. We will see that the frequency (or wavelength) of the electromagnetic field dictates the type of interaction that can occur with a given material: for example, microwaves can be absorbed or emitted by molecules via their rotational modes, while gamma rays (which have much higher frequency or energy) are generated and absorbed through nuclear interactions. In this chapter, we aim to briefly review the most common electromagnetic regions and their applications.

One striking feature of the visible spectrum is that it occupies a very narrow range of the electromagnetic spectrum (figure 1.1). The lowest frequency detectable by our retina ($\simeq$760 nm wavelength, dark red color) only differs by a factor of two from the highest detectable frequency ($\simeq$380 nm, violet color). By comparison, our sensitivity to sound waves spans three orders of magnitude in frequency, 20 Hz–20 kHz. Yet, the visual system is by far the most complex and valuable of the human senses. We will see in chapter 6 that the visible range of the spectrum matches the peak of the black body radiation of the Sun. Thus, evolving to develop eye sensitivity in the visible range provides an evolutionary advantage.

Next, we will describe the common frequency intervals of the electromagnetic spectrum. To separate these regions, we use the frequency (the number of field oscillations per second, ν) and, equivalently, the wavelength (the distance traveled by the field during one period, $\lambda = c/\nu$, c the speed of light).

1.1.1 Gamma rays

Gamma rays define the highest frequency, highest energy photons known to man ($E = h\nu = hc/\lambda$, with h Planck's constant). The wavelength is $\lambda < 0.01$ nm and has

doi:10.1088/978-0-7503-1644-6ch1

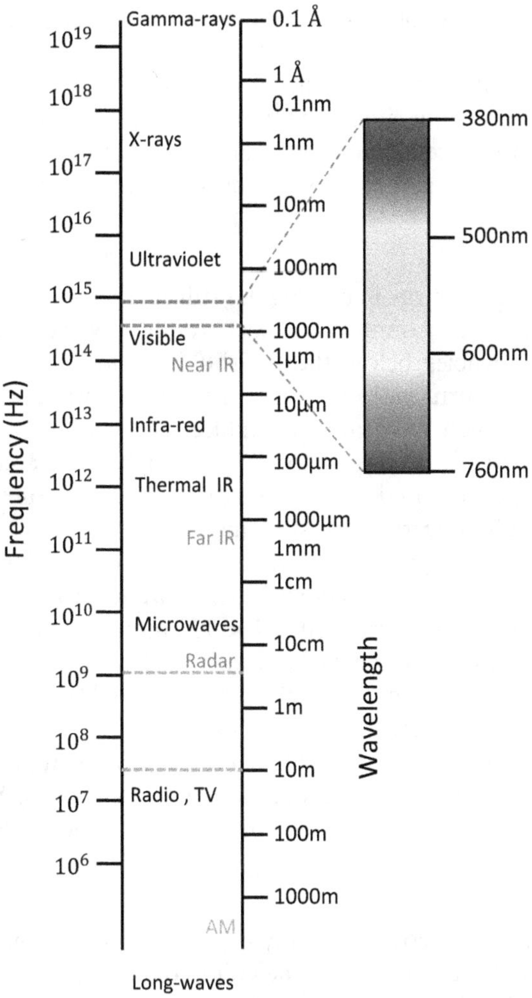

Figure 1.1. The electromagnetic spectrum, as a function of frequency and wavelength, as indicated.

no lower limit. They are produced, and thus used to study, high energy phenomena in nuclear physics. Gamma rays are also used in astronomy as they can report on high-energy interaction deep into the stars.

In medicine, gamma rays are commonly used for diagnostic imaging, for example, in positron emission tomography. In recent years, gamma radiation has been increasingly used for cancer radiation therapy.

1.1.2 X-rays

X-ray radiation has lower wavelengths, but is still capable of penetrating large objects without significant absorption. Telescopes based on x-ray radiation are broadly used for studying high-energy processes in, for example, black holes, neutron stars, and coronas of stars. Gamma and x-ray telescopes must operate in space due to the high absorption in the atmosphere.

X-rays are routinely used in medicine for radiography and computed tomography. This radiation is not significantly absorbed in soft tissues (without contrast agents), but provides excellent contrast of bone. X-ray wavelengths are above λ 10 nm and overlap with gamma rays at the short frequency end.

1.1.3 Ultraviolet

Ultraviolet (UV) is below x-ray in energy, but still energetic enough that it can ionize atoms, as can x-rays and gamma rays. As a result, UV can be harmful to living tissue. Even at frequencies below the threshold for ionization, UV can damage chemical bonds and turn molecules into reactive compounds, which are also potentially harmful. Such radiation can induce, for example, irreparable damage to the DNA of living cells. Exposure to UV is strongly correlated with skin cancer. Because UV is commonly used as the excitation light in fluorescence microscopy, cell viability is often a concern when studying live specimens.

UV wavelength is roughly within the range $10 < \lambda < 400$ nm. Vacuum UV (VUV) defines the spectral range below 200 nm, while extreme UV (EUV) covers the end of the high energy spectrum, $10 < \lambda < 121$ nm. The longer wavelength range is separated into UVA (315–400 nm), UVB (280–315 nm), and UVC (100–280 nm).

The atmosphere blocks approximately 75%–80% of UV radiation. The portion that reaches the surface of the Earth consists of >95% UVA, <5% UVB, and no UVC. The sunlight reaching the Earth consists only of approximately 3% UV. Despite its potential harmful effects to living organisms, UVB is involved in metabolizing vitamin D, which is crucial for developing bone strength.

1.1.4 Visible

The visible (VIS) spectrum covers roughly the 380–760 nm range, where the human retina exhibits its highest sensitivity. The colors of surrounding objects are due to their absorption and reflection properties. For example, plants look green because chlorophyll absorbs strongly in blue and red. Thus, they reflect the 'left-over light', the green. A prism separates the sunlight into colors because the glass has *dispersion*, i.e. its refractive index depends on wavelength, and the refraction angle at the prism surface is different for different colors. As described later, in chapter 4, the human retina reaches a maximum sensitivity around 550 nm, corresponding to the green color.

1.1.5 Infrared

Infrared (IR) covers a broad range of wavelengths that are larger than that of the red color, 0.75–1.000 μm. This range is further split into three regions: near infrared (NIR, 0.75–2.5 μm) mid-infrared (MIR, 2.5–10 μm), and far-infrared (FIR, 10–100 μm). NIR radiation is involved in similar electronic charge interactions as visible light and can be detected by solid-state sensors (see chapters 12–13). MIR is a spectral range in which bodies at room temperature can radiate. Because MIR contains fine spectral structures due to specific molecular vibrations, it is sometimes called the *fingerprint* region. The FIR region contains radiation absorbed (and generated) by the rotational modes of molecules. For example, water molecules in the atmosphere absorb strongly in this range and render the atmosphere opaque to these frequencies, except for $\lambda > 200\ \mu m$. The NIR spectral region is broadly employed in biophotonics, particularly due to the tissue penetration, which is typically deeper that in the VIS range. IR is also used to study tissues with chemical specificity.

1.1.6 Terahertz

Terahertz (THz) (also submillimeter) radiation covers the wavelength region 0.1–1 mm. Due to the significant absorption by the atmosphere, THz cannot be used for long-range communications. With recent improvements in sources and detectors, THz radiation has received significant attention of late for imaging applications.

1.1.7 Microwaves

Microwaves span the wavelengths $1\ mm < \lambda < 100\ mm$. They are typically generated by antennas (e.g. in cell phones). However, microwaves can interact with rotational and vibrational modes of molecules, eventually converting the energy into heat. Heating food by microwave ovens is a notorious example, where energy is transferred to the rotational modes of water and, finally, into heat. The microwave spectrum is broadly used in satellite communication, radar, and wireless connections. Microwaves have been used for the thermal treatment of tumors.

1.1.8 Radiowaves

Radio frequency waves are emitted and detected by antennas of various dimensions, depending on the wavelength. They cover the broad range of wavelengths $\lambda > 10\ cm$. Radiowaves are by far the most commonly used region of the electromagnetic spectrum for long distance communication: radio, television, global positioning systems, radar, etc. As a result, the frequency bands are highly regulated.

Biophotonics encompasses the interaction of tissue with the electromagnetic fields, generally, from IR to UV. The type of interaction between the field and specimen determines what information can be extracted from the tissue. The subsequent volumes discuss in detail the principles of various methods that have a particular field–matter interaction as the starting point. The particular frequency range used determines whether the object of interest is:

- dispersive, that is, with a wavelength-dependent refractive index;
- spatially inhomogeneous, that is, with a spatially-dependent refractive index;
- anisotropic, that is, with a polarization-dependent refractive index;
- nonlinear, that is, with an irradiance-dependent refractive index.

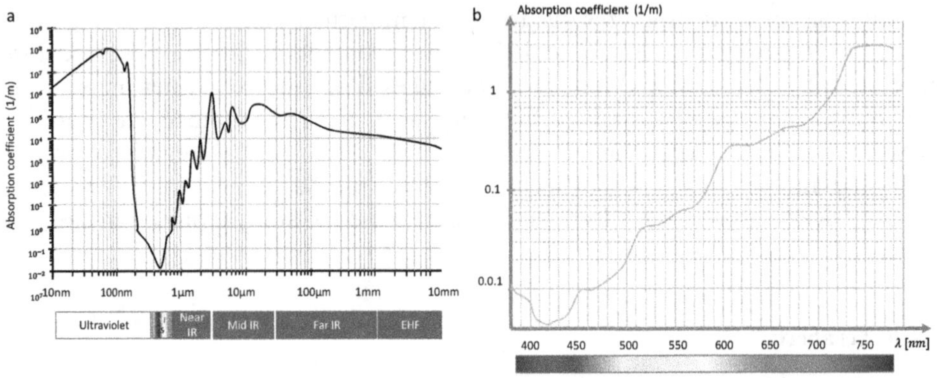

Figure 1.2. a) Absorption coefficient of water over a broad range of the electromagnetic spectrum (log–log scale). b) Water absorption across the visible spectrum (semi-log *y* plot).

For now, we postpone the rigorous discussion of these phenomena. Instead, we briefly describe next the spectral absorption of water and hemoglobin, as they are crucial contributors to the overall light attenuation in tissues.

1.2 Spectral absorption of water

Water is the dominant component of most tissues. Therefore, the absorption coefficient of water at various frequencies plays an important role in the overall attenuation of the field [1]. Figure 1.2 illustrates the water coefficient of absorption versus wavelength. It can be seen that the absorption is very high in UV and most of IR, and low around the visible region. This should not be surprising, as water looks transparent to the eye. Across the visible spectrum, we observe a monotonous increase in absorption from blue to red (figure 1.2(b)). Although we might be tempted to conclude that blue light should always be used for deep tissue imaging, this conclusion would be premature. First, there are other absorptive molecules in the tissue (e.g., hemoglobin, discussed in section 1.3) that have to be considered. Second, and perhaps, most importantly, scattering, which exhibits a strong wavelength dependence, has a significant effect on the overall attenuation.

1.3 Spectral absorption of hemoglobin

Hemoglobin is the most abundant protein in the red blood cell. As such, its spectral absorption [2] has to be considered when performing experiments on tissue, particularly when aiming for deep imaging. As shown in figure 1.3, there are differences in the absorption spectrum for the oxygenated versus deoxygenated hemoglobin.

However, both curves exhibit a strong absorption peak in the blue region of the visible spectrum, followed by a low peak in the green region. This, of course, explains the particular red color of blood. Tuning the wavelength across the visible spectrum, for example at points where the oxy-and deoxy-hemoglobin absorb (1) equally well, and (2) very differently, allows the investigator to infer the ratio of the

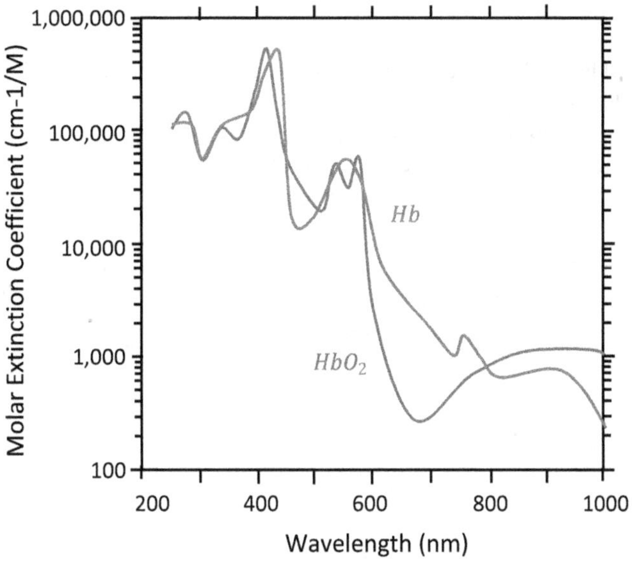

Figure 1.3. Molar absorption coefficient for deoxygenated (red) and oxygenated (blue) hemoglobin.

two species, which can reveal information about the rate of metabolic processes in the tissue.

The strong absorption in the blue region overwhelms the relative transparency of water in the same region. Accounting for the scattering effects, which become less significant at longer wavelengths, yields that the deepest tissue penetration can actually be obtained using NIR radiation. More on this can be found in the subsequent volumes.

1.4 Problems

1. Choose a typical frequency corresponding to radiowaves, microwaves, THz, IR, VIS, UV, x-rays, and gamma rays, and convert it into wavelength and then into photon energy (units of eV).
2. A field for irradiance $I_0 = 1\ \text{W/m}^2$ is incident on a 10 m thick aquarium filled with water. Using the plots in figure 1.2, estimate the irradiance emerging from the aquarium, if the electromagnetic field has the following wavelength
   a) $\lambda = 50$ nm
   b) $\lambda = 500$ nm
   c) $\lambda = 10$ μm
   d) $\lambda = 1$ mm.
3. What is the percentage of irradiance loss due to Hb absorption in a single red blood cell of thickness $t = 2.5$ μm, if the wavelength is (use figure 1.3)
   a) $\lambda = 400$ nm
   b) $\lambda = 500$ nm
   c) $\lambda = 600$ nm?

References

[1] Curcio J A and Petty C C 1951 The near infrared absorption spectrum of liquid water *J. Opt. Soc. Am.* **41** 302–4

[2] Horecker B L 1943 The absorption spectra of hemoglobin and its derivatives in the visible and near infra-red regions *J. Biol. Chem.* **148** 173–83

IOP Publishing

Principles of Biophotonics, Volume 2
Light emission, detection, and statistics
Gabriel Popescu

Chapter 2

Radiometric properties of light

Radiometry measures properties of the electromagnetic field as revealed by its interaction with a photodetector [1, 2]. We gain experimental knowledge about the light properties by detecting these radiometric quantities. *Photometry*, on the other hand, discussed in chapter 4, deals with quantities revealed by the interaction with the human eye. Thus, radiometry provides an objective measure of the optical field, while photometry accounts for the physiological response of the human retina. Generally, in scientific studies that aim to unravel the properties of light in an absolute, unbiased sense, radiometric quantities must be used. However, when the problem at hand involves the human perception, such as in the case of consumer cameras, displays, etc, *photometric* quantities are used instead. The following quantities of an electromagnetic field are most commonly used in radiometry and will be encountered throughout the book.

2.1 Energy

The energy of an optical field quantifies its ability to perform work. Examples include accelerating particles, heating materials, generating photoelectrons in a detector, etc. Energy is measured in *joules* (J),

$$\begin{aligned} Q &= \text{energy} \\ [Q] &= \text{J}. \end{aligned} \tag{2.1}$$

Light is generated by a process of energy conversion, for example, from thermal to optical energy in the case of an incandescent bulb, from optical to optical energy in the case of an optically-pumped laser, from electrical to optical energy in the case of an electrically-pumped laser, from chemical to optical energy in the case of fireflies (and chemically-excited lasers), etc. An optical field can exchange energy with various materials, in which case the interaction is called *inelastic*. Another class of

doi:10.1088/978-0-7503-1644-6ch2

light–matter interactions are *elastic,* when the light can change direction of propagation (or momentum), without loss (or gain) in energy.

2.2 Energy density

Energy is generally defined over a certain (finite) volume in space, V (figure 2.1). Thus, we can introduce an energy density, W, as

$$\begin{aligned} W &= \frac{Q}{V} \\ [W] &= \frac{\mathrm{J}}{\mathrm{m}^3}. \end{aligned} \tag{2.2}$$

The energy density is a necessary quantity for describing how a given amount of energy concentrates in space: the same 1 J of energy will likely produce very different outcomes when distributed over 1 m^3 versus over 1 $\mu\mathrm{m}^3$ (an 18-order of magnitude difference in energy density).

2.3 Power

Power (denoted by P) represents energy exchanged (i.e., emitted, absorbed, detected, reflected, transmitted, scattered, etc) per unit of time (see figure 2.2). The unit of power is J/s or Watt (W),

$$\begin{aligned} P &= \frac{dQ(t)}{dt} \\ [P] &= \mathrm{J/s} \\ &= \mathrm{W}. \end{aligned} \tag{2.3}$$

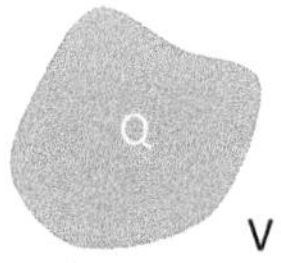

Figure 2.1. Energy density is the ratio of energy Q to the volume V, $W = Q/V$.

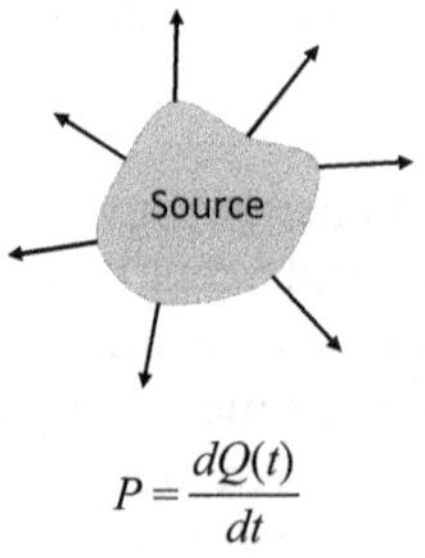

Figure 2.2. Power radiated by a source is the energy delivered per unit of time.

The power of an electromagnetic field is useful in describing how a certain amount of energy is concentrated in time. For example, 1 mJ of energy may not seem significant at first, but when delivered in a pulse of 25 fs long (1 fs = 10^{-15} s), yields P = 40 GW (4 × 10^{10} W), sufficient to produce damage in many materials. This output is common for commercially available femtosecond lasers. Interestingly, the largest nuclear plant in the US, the Palo Verde in Arizona, generates an order of magnitude less power, ~4 GW. It means that, for a very short period of time, the laser delivers more energy than the nuclear plant. Of course, in the long run, the power plant outputs a lot more energy than the laser (otherwise, we would need several nuclear plants to power the laser!). The reason for this is that the laser pulses are emitted rather sparsely, typically, every microsecond, which means that, for the most part, the laser yields no energy. For example, in one second, the laser outputs 10^6 pulses, adding up to 1 kJ, while the power plant yields 4 GJ, a factor of a million more. Thus, for pulsed lasers, it is necessary to distinguish between the peak power, 40 GW in this case, and its average power, 1 kJ s^{-1} = 1 kW, which can be very different in value.

2.4 Temporal power spectrum

The power spectrum (or spectral power), represents the power per unit of optical frequency, ν (see figure 2.3).

$$S(\nu) = \frac{dP(\nu)}{d\nu} \qquad [S(\nu)] = \text{W/Hz}. \tag{2.4}$$

Note that the optical frequency, ν, in Hz, differs from the angular frequency, ω, in rad/s, by a factor of 2π, namely, $\omega = 2\pi\nu$. We chose to express the Fourier transform of a signal f(t) (see volume 1) in terms of angular frequencies (ω), $\tilde{f}(\omega)$, as it removes the 2π factors from the exponents. However, for spectral densities, we often use ν,

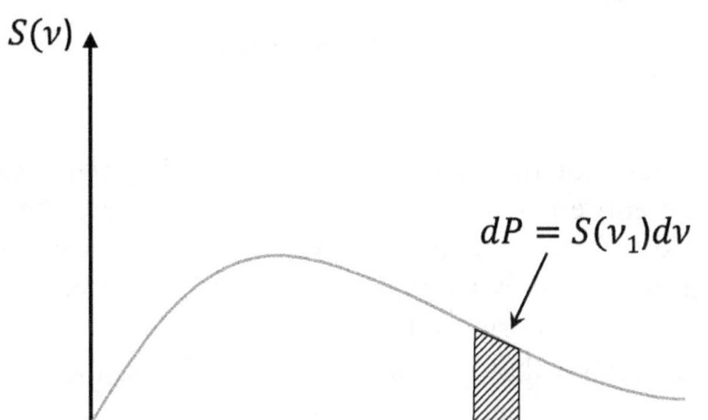

Figure 2.3. Temporal power spectrum. The infinitesimal area indicated is the power contained in the vicinity of ν_1.

leading to units of W/Hz, rather than the more cumbersome, W/(rad/s). Of course, the power spectrum expressed in terms of angular frequency, $S(\omega)$ (see, e.g., equation (4.26), volume 1), is an entirely equivalent representation, and is related to $S(\nu)$ simply by (we use the same symbol, S, but carry the argument so there no confusion as to which spectral representation we use),

$$\begin{aligned} S(\omega) &= \frac{dP}{d\omega} \\ &= \frac{dP}{d\nu}\frac{d\nu}{d\omega} \\ &= \frac{1}{2\pi}S(\nu). \end{aligned} \tag{2.5}$$

The power spectrum of an optical field describes how the power is distributed with respect to the frequency content. For example, for the same power level, the light from the Sun is distributed over a much broader frequency range than, say, that from a stabilized laser.

The total power is obtained by integrating the power spectrum,

$$P = \int_0^{\infty} S(\nu)d\nu. \tag{2.6}$$

From equation (2.6), we see that Parseval's theorem, discussed in Volume 1, section 4.3 is a manifestation of energy (or power) conservation. Thus, for a signal f(t) and its Fourier transform, $\tilde{f}(\omega)$, Parseval's theorem states (see equation (4.15a), Volume 1)

$$\begin{aligned} \int_{-\infty}^{\infty} |f(t)|^2 dt &= \frac{1}{2\pi}\int_{-\infty}^{\infty} |\tilde{f}(\omega)|^2 d\omega \\ &= \frac{1}{2\pi}\int_{-\infty}^{\infty} S(\omega)d\omega \\ &= \int_{-\infty}^{\infty} S(2\pi\nu)d\nu \\ &= 2P. \end{aligned} \tag{2.7}$$

Equation (2.7) states that the total power of a signal is the same whether it is integrated over its entire spread in time or frequency. Note that the integral in equation (2.7) yields $2P$ rather than P, because the integral also spans the negative frequencies (equation (2.6) only integrates over the positive frequencies). For a real signal, f, its Fourier transform is Hermitian, $\tilde{f}(-\omega) = \tilde{f}^*(\omega)$ (see section 7.2, Volume 1), which implies that $|\tilde{f}(-\omega)|^2 = |\tilde{f}(\omega)|^2$. Thus, the total power in the negative frequencies is the same as in the positive frequencies, which is why equations (2.6) and (2.7) differ by a factor of two.

Often, experimentally, we measure the power spectrum with respect to the wavelength, $\lambda = c/\nu$, with c the speed of light in a vacuum. We can write the wavelength power spectrum as

$$\begin{aligned} S(\lambda) &= \frac{dP}{d\lambda} \\ &= \frac{dP}{d\nu} \cdot \frac{d\nu}{d\lambda} \\ &= -\frac{c}{\lambda^2} S(\nu). \end{aligned} \tag{2.8}$$

Note that the unit changes accordingly, in terms of power per unit wavelength, $[S(\lambda)] = \text{W/m} = 10^{-6}\ \text{W}/\mu\text{m}$.

The power spectrum is a non-negative quantity and, thus, can be normalized to a probability density, as follows,

$$s(\nu) = \frac{S(\nu)}{\int_0^\infty S(\nu) d\nu}. \tag{2.9}$$

The meaning of the normalized quantity $s(\nu)$ is the probability of having radiation in the infinitesimal frequency interval $(\nu, \nu + d\nu)$. We can define the first two moments of this distribution, which yield the mean frequency, and the frequency standard deviation (bandwidth),

$$\nu_0 = \int_0^\infty \nu s(\nu) d\nu \tag{2.10a}$$

$$\sigma_\nu^2 = \int_0^\infty \nu^2 s(\nu) d\nu - \nu_0^2. \tag{2.10b}$$

In equations (2.10*a*) and (2.10*b*), ν_0 is the mean frequency and σ_ν^2 the variance. The standard deviation, σ_ν, is a physical measure for the frequency bandwidth (see figure 2.4).

2.5 Intensity: spatial power spectrum

Intensity represents the power per unit solid angle, Ω (figure 2.5(a)),

$$\begin{aligned} \mathrm{I}_\Omega &= \frac{dP(\Omega)}{d\Omega} \\ [\mathrm{I}_\Omega] &= \text{W/sterradian} \\ &= \text{W/srad}. \end{aligned} \tag{2.11}$$

The infinitesimal solid angle, $d\Omega$, is defined as

$$\begin{aligned} d\Omega &= \frac{\hat{\mathbf{r}} \cdot d\mathbf{A}}{r^2} \\ &= \hat{\mathbf{r}} \cdot \hat{\mathbf{n}} \frac{dA}{r^2} \\ &= \sin\theta\ d\theta\ d\phi, \end{aligned} \tag{2.12}$$

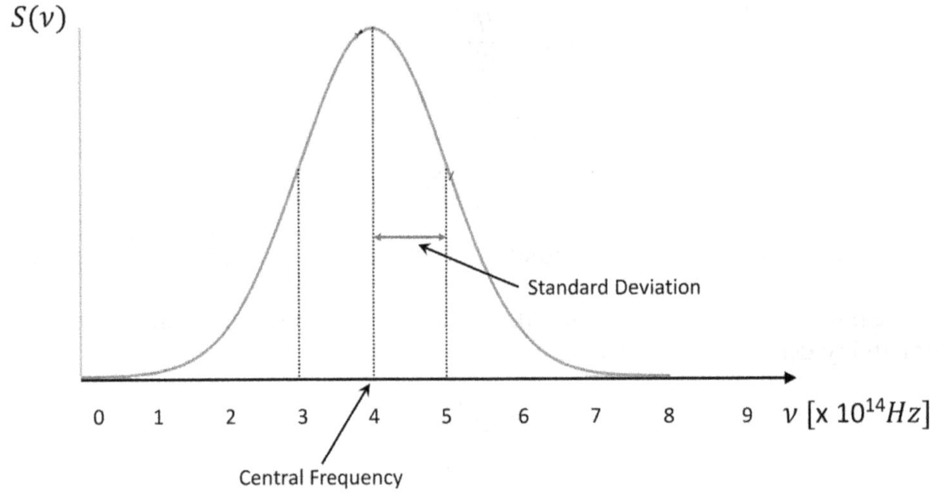

Figure 2.4. Optical spectrum with central frequency and standard deviation defined as first and second order moments of the frequency distribution.

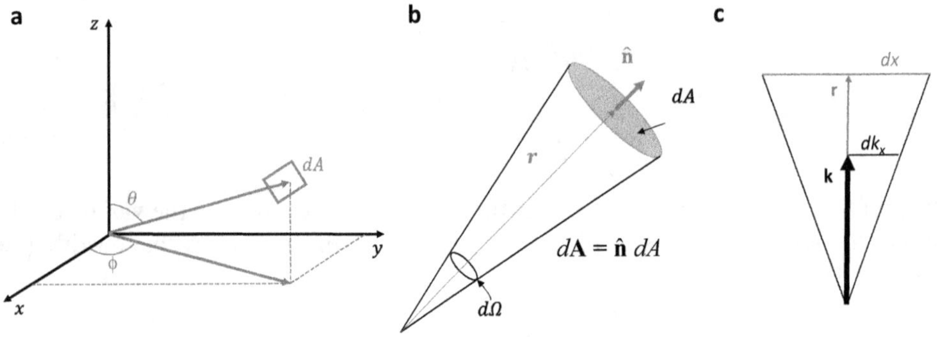

Figure 2.5. a) Representation of an infinitesimal area in the 3D space. b) Infinitesimal solid angle. c) Illustration of how the solid angle can be expressed in terms of wavevector.

where $\hat{\mathbf{r}}$ is the unit position vector and dA the infinitesimal area vector, with the direction along the normal at the surface, $\hat{\mathbf{n}}$, $d\mathbf{A} = \hat{\mathbf{n}}\, dA$ (see figure 2.5(b)).

The solid angle subscribed by a certain area at a distance $\mathbf{r}$ with respect to a point of reference is obtained by the integration

$$\begin{aligned}\Omega &= \int_A \frac{\hat{\mathbf{r}} \cdot d\mathbf{A}}{r^2} \\ &= \int_{\phi_1}^{\phi_2} \int_{\theta_1}^{\theta_2} \sin\theta \, d\theta \, d\phi,\end{aligned} \tag{2.13}$$

where $\theta_{1,2}$ are the polar and $\phi_{1,2}$ the azimuthal angle limits as subtended by the surface. Intensity is a *directional* quantity, that is, it depends on the particular direction of interest. Special sources that emit constant intensity in all directions are called *isotropic emitters.*

It is important to realize that the solid angle can be expressed in terms of wavevectors, **k**, and that the intensity is related to the *spatial power spectrum.* Considering the solid angle along the z-axis for simplicity (figure 2.5(c)), we see that it can be expressed in two different ways

$$\begin{aligned} d\Omega &= \frac{\hat{\mathbf{r}} \cdot d\mathbf{A}}{r^2} \\ &= \frac{dx\,dy}{r^2} \\ &= \frac{dk_x\,dk_y}{k^2}, \end{aligned} \tag{2.14}$$

where k is the magnitude of the wavevector ($k = \omega/c$ in air), and dk_x, dk_y the infinitesimal variation of the **k**-vector in two perpendicular directions, $d^2\mathbf{k}_\perp$. In other words, the intensity can be interpreted as a distribution of power over transverse spatial frequencies,

$$\begin{aligned} \mathrm{I}_\Omega &= k^2 \frac{d^2 P}{d\mathbf{k}_\perp{}^2} \\ &= k^2 S(\mathbf{k}_\perp). \end{aligned} \tag{2.15}$$

In equation (2.15), $\mathbf{k}_\perp = (k_x, k_y)$ and $S(\mathbf{k}_\perp)$ is the 2D spatial power spectrum that describes the power delivered in the vicinity of the given **k**-vector. Up to the k^2 constant, the intensity is equal to this spatial power spectrum. Therefore, the *intensity* is the *spatial* analog of the temporal *power spectrum* described in section 2.4. Analog to the temporal frequency domain, the 2D *spatial* power spectrum can be obtained from the 2D Fourier transform of the field, $S(\mathbf{k}_\perp) = |\tilde{U}(\mathbf{k}_\perp)|^2$ (see chapter 5, volume 1 for a description of the 2D Fourier transform and its properties).

2.6 Irradiance

Irradiance is the power delivered per unit area,

$$\begin{aligned} I &= \frac{dP}{dA} \\ [I] &= \frac{\mathrm{W}}{\mathrm{m}^2}. \end{aligned} \tag{2.16}$$

This quantity is encountered often, when assessing how a certain amount of power is distributed spatially in a plane. For example, the irradiance of the visible light from the Sun at the surface of the Earth is approximately 10^3 W m^{-2} (on an average clear day, at sea level). In microscopy, at the same power of the illumination light, a confocal geometry generates much higher irradiance than a wide-field set-up (see figure 2.6). A wide-field microscope may spread the illumination light over, say,

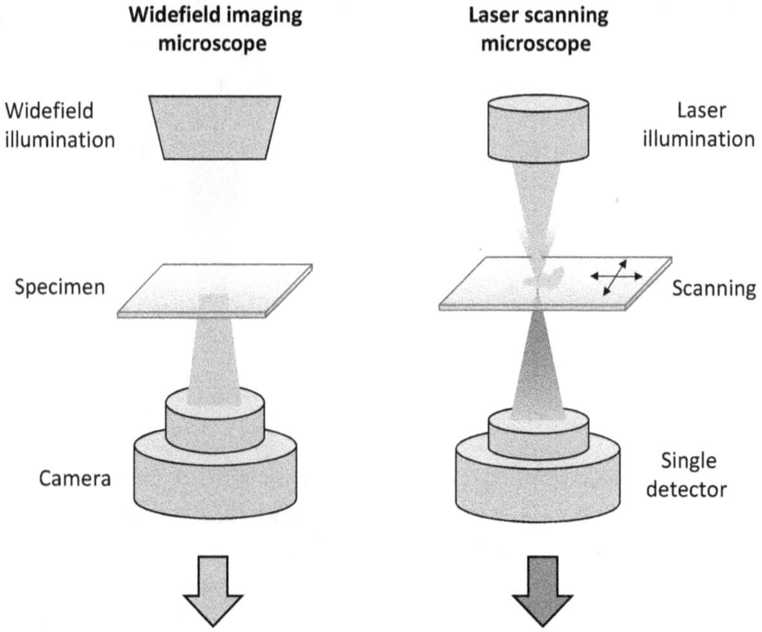

Figure 2.6. Wide-field vs confocal illumination. At the same total power, the confocal illumination delivers orders of magnitude higher *irradiance*.

a 1 mm diameter beam, while in a confocal instrument, the same light can be focused to a 1 μm diameter spot. Thus, the confocal microscope will deliver a factor of $(1\ \text{mm}/1\ \mu\text{m})^2 = 10^6$ higher irradiance.

2.7 Spectral irradiance

The spectral irradiance describes the power per unit area and unit of optical frequency,

$$\begin{aligned} I_\nu &= \frac{d^2P}{dA\, d\nu} \\ [I] &= \frac{\text{W}}{\text{m}^2\ \text{Hz}}. \end{aligned} \tag{2.17}$$

For example, I_ν can be used to describe how the 1 KW m^{-2} irradiance from the Sun is distributed over the visible spectrum. The spectral irradiance can be expressed in terms of angular frequency interval, I_w, or wavelength, I_λ, following the procedure described in section 2.4.

2.8 Radiance

Radiance represents the power emitted or absorbed by a surface, per unit solid angle, per unit projected area,

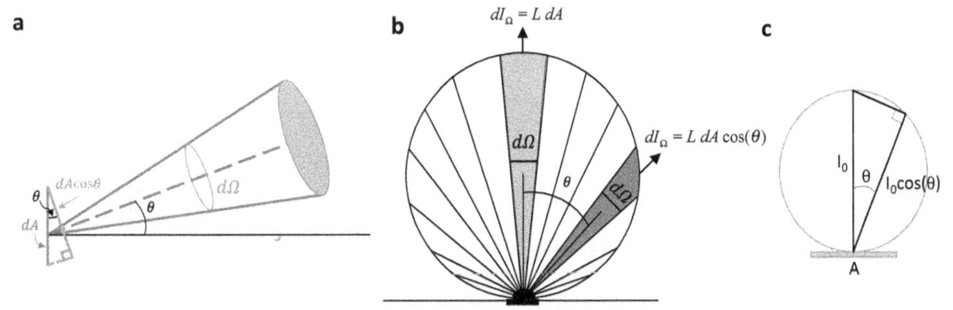

Figure 2.7. a) The cosine term in equation 2.18 is due to the projection of the source normal onto the direction of interest. b) Irradiance from a Lambertian in a certain direction source. c) Irradiance from a Lambertian source describes a circle tangent at the source.

$$
\begin{aligned}
L &= \frac{d^2P(\Omega, A)}{d\Omega\, dA \cos\theta} \\
[L] &= \frac{\mathrm{W}}{\mathrm{m}^2\ \mathrm{srad}}.
\end{aligned}
\tag{2.18}
$$

The $\cos\theta$ term in equation (2.18) is due to the projection of the source normal onto the direction of interest (figure 2.7(a)). Note that radiance is a directional quantity, similar to intensity. A particular type of sources have the property that L is *constant*, that is, it does not depend on the observation angle or the particular region on the source. These sources are called *Lambertian*. A constant radiance with respect to angle implies that the intensity, $I_\Omega = \frac{dP}{d\Omega}$, has a $\cos\theta$ dependence on angle, as the intensity emitted by the element of area dA is

$$
dI_\Omega = L\ dA\cos(\theta). \tag{2.19}
$$

Equation (2.19) is sometimes referred to as *Lambert's cosine law*. The intensity versus angle can be represented on a circle, with the source tangent at the circle, as the angle that subtends the diameter (normal intensity) is always 90 degrees (figure 2.7(b)–(c)). Interestingly, when light propagates through a strongly scattering medium, such as a cloud or diffuser, that medium becomes a Lambertian (secondary) source. Most projector screens used in conference rooms are designed to diffuse light and, thus act as Lambertian sources.

2.8.1 Radiance conservation theorem

It can be shown that the radiance through a lossless optical system is *conserved*. This important statement is just another manifestation of the energy conservation and can be proven using figure 2.8. We start by expressing the power of the radiation in two ways corresponding to the two different surfaces, dA_1, dA_2:

$$
\begin{aligned}
dP &= L_1\, d\Omega_1\, dA_1 \cos\theta_1 \\
&= L_2\, d\Omega_2\, dA_2 \cos\theta_2.
\end{aligned}
\tag{2.20}
$$

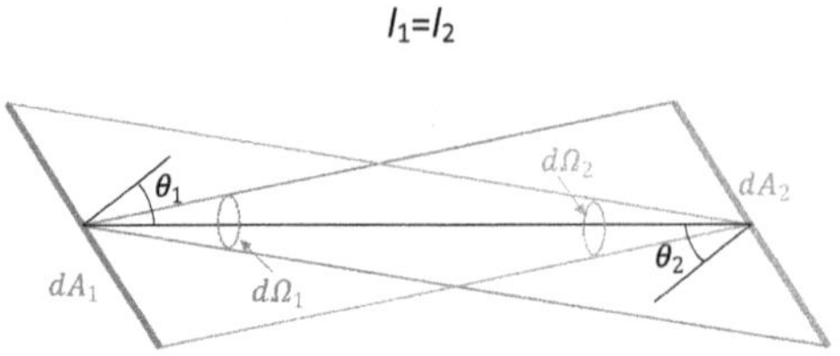

Figure 2.8. Conservation of radiance.

Note that $d\Omega_1$ is the solid angle subtended by dA_2 as seen from dA_1 and $d\Omega_2$ is subtended by dA_1 as seen from dA_2, thus

$$d\Omega_1 = \frac{dA_2 \cos\theta_2}{r^2} \tag{2.21a}$$

$$d\Omega_2 = \frac{dA_1 \cos\theta_1}{r^2}. \tag{2.21b}$$

Combining equations (2.20) and (2.21) leads to $L_1 = L_2$ (*q.e.d.*).

This conservation theorem has practical implications in designing optical systems. One implication is that the light can only be spread over a larger area if the angular distribution narrows down accordingly (and *vice versa*). For example, in a microscope, the field of view at the sample plane is magnified by a significant factor at the image plane. This implies that the angular (spatial frequency) distribution at the image plane is narrower by the same factor. Sometimes the radiance is *not* conserved in a lossy optical system, due to, for example, *vignetting*, a method of obscuring the optical path developed for artistic effects in photography, but undesirable in most scientific experiments. Some references use the French word 'etendue' to describe radiance and 'conservation of etendue' as the respective theorem.

As described in section 2.5, the distribution of power with respect to solid angles (*intensity*) is, up to a constant, the spatial power spectrum. Thus, radiance can be expressed as a distribution in both space and spatial frequency,

$$\begin{aligned} L &= k^2 \frac{d^4P}{d\mathbf{k}_\perp{}^2\, dA \cos(\theta)} \\ &= k^2 \frac{d^4P}{(dx\, dk_x)(dy\, dk_y)}, \end{aligned} \tag{2.22}$$

where dx, dy are the infinitesimal spatial intervals normal to the direction of interest, as shown in figure 2.5(c). Thus, conservation of radiance expresses that, for a given field, the space–bandwidth product is constant,

$$\begin{aligned} dx\, dk_x &= \text{const} \\ dy\, dk_y &= \text{const}, \end{aligned} \tag{2.23}$$

where dk_x, dx, dk_y, dy can be considered as the power-weighted spreads in space and spatial frequency along x and y. Note the similarity between equations (2.23*a*) and (2.23*b*) and the uncertainty relation that states (equation (8.5), Volume 1)

$$dx\, dk_x \geqslant 1/2. \tag{2.24}$$

While the uncertainty relation establishes that the spreads in space and spatial frequency cannot both be arbitrarily small and reaches a minimum of ½ for the Gaussian distribution, the radiance conservation states that, for a given field distribution, the product of the spreads remains constant.

2.9 Spectral radiance

Spectral radiance is defined as the radiance per unit of frequency,

$$\begin{aligned} L_\nu &= \frac{dL}{d\nu} \\ [L_\nu] &= \frac{\text{W}}{\text{m}^2\,\text{srad Hz}}. \end{aligned} \tag{2.25}$$

In essence, L_ν allows us to characterize the emission from a surface element into a certain direction, like the radiance, but with the added information about the temporal frequency content. Similar to all the previous spectral quantities, we can define spectral radiance with respect to angular frequency, L_ω, and wavelength, L_λ, by changing the variable and using the respective Jacobian (equations (2.5) and (2.8)).

2.10 Exitance

Radiant exitance is defined as the amount of power emitted by a source per unit area

$$\begin{aligned} M &= \frac{dP}{dA_s} \\ [M] &= \frac{\text{W}}{\text{m}^2}. \end{aligned} \tag{2.26}$$

Note that, unlike *irradiance,* which shares the same units, exitance is the property of a source, while irradiance is the property of the field. Exitance represents the radiance of the source integrated over all solid angles

$$M = \int_\Omega L(\Omega)d\Omega. \tag{2.27}$$

2.11 Spectral exitance

Similar to the spectral irradiance, spectral exitance is defined as the exitance per unit frequency, or power emitted by the source per unit area, per unit frequency

$$
\begin{aligned}
M_\nu &= \frac{dM}{d\nu} \\
&= \frac{d^2P}{dA_s\, d\nu} \qquad (2.28) \\
[M_\nu] &= \frac{\mathrm{W}}{\mathrm{m}^2\ \mathrm{Hz}}.
\end{aligned}
$$

2.12 Problems

1. A source of area $A_S = 1$ cm^2 has a radiance of $L = 10$ W/m^2 srad (figure 2.9).
   a) How much power does a lens collect, given that it has a diameter of $D = 2.5$ cm^2 and is placed at a distance $L = 100$ cm from the source, parallel to the source and centered on the same axis?
   b) How much power does the lens collect if it is shifted off-axis by $d = 1$ cm?
2. Consider a Lambertian source, of radiance L and area A. Plot in *polar coordinates* the radiance and intensity within the angular range, $\theta \in (-\pi/2, \pi/2)$ (figure 2.10).

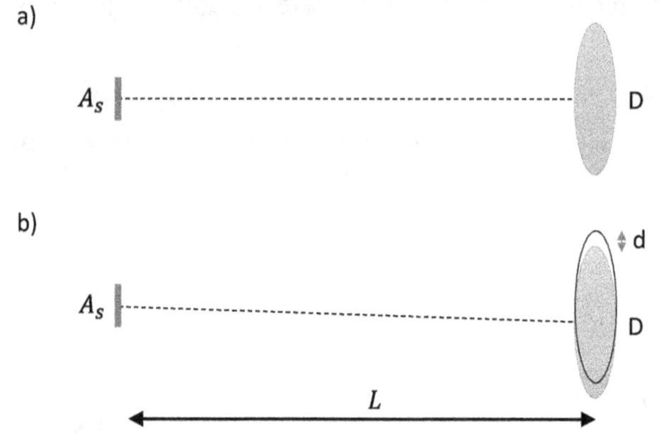

Figure 2.9. Problem 2.1: a) centered lens, b) off-centered lens.

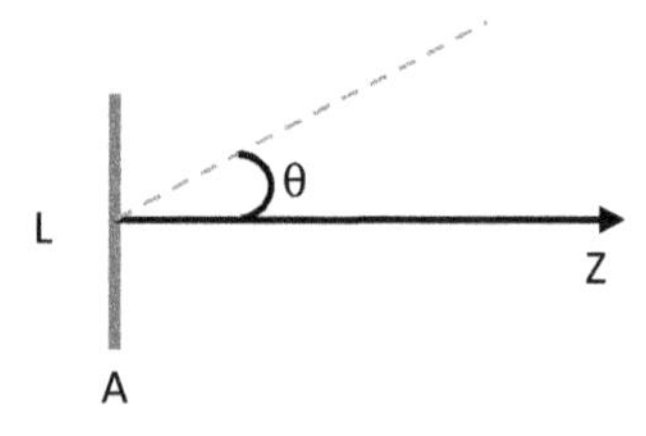

Figure 2.10. Problem 2.2.

3. A point source is characterized by an intensity $I_\Omega = 1$ W/srad. An optical system placed 100 m away has an aperture of 1 m^2 and focuses the incident light to a spot of 1mm diameter (figure 2.11).
   a) What is the irradiance, I, in W/m^2, at the focal plane?
   b) Ignoring aberrations and other effects, how does the result change if the point source is shifted in the z-direction by $\Delta z = 1$ m, away from the optical system (figure 2.11)?
   c) How does the result change if the point source is shifted in x by $\Delta x = 1$ m?
4. Given the flat Lambertian source shown in figure 2.10, of radiance $l = 1$ kW/m^2 srad:
   a) What is the intensity along the direction shown ($\theta = 30°$ with respect to the normal)?
   b) What is the irradiance at a distance $L = 1$ m along that direction?
5. A non-Lambertian source is characterized by a radiance $l(\theta) = l_1 \cos^2 \theta$. The source detector geometry is illustrated in figure 2.12 ($l_1 = 1$ W/m^2 × srad).
   a) What is the intensity at the detector?
   b) What is the irradiance at the detector?
   c) What is the total power falling on the detector?

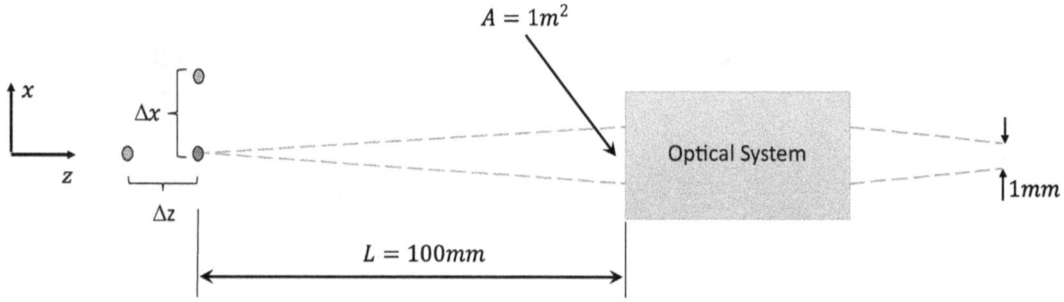

Figure 2.11. Problem 2.3.

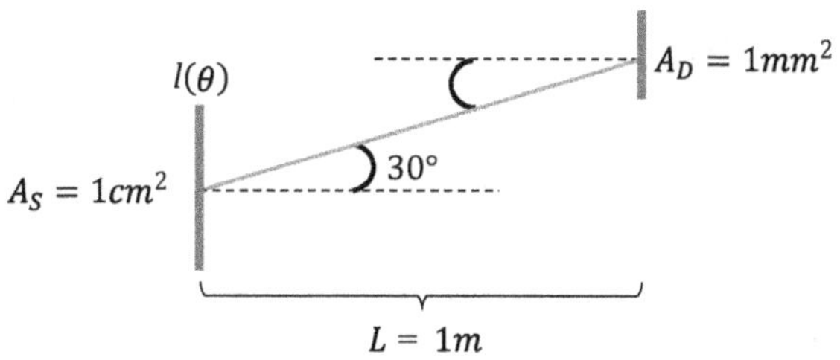

Figure 2.12. Problem 2.5.

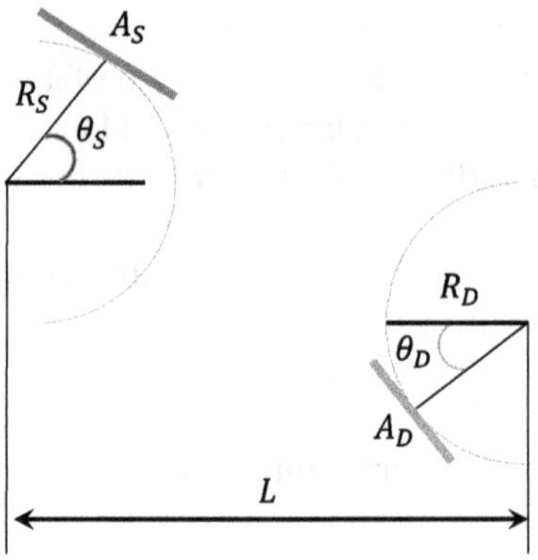

Figure 2.13. Problem 2.6.

6. A Lambertian source of area A_s, and a detector of area A_D, are scanned along circular trajectories, of radii R_S, R_D (figure 2.13). If the distance between the centers of the two circles is L, compute the power falling on the detector as a function of θ_S, θ_D.

References

[1] Boyd R W 1983 *Radiometry and the Detection of Optical Radiation* (Wiley Series in Pure and Applied Optics) (New York: Wiley), vii p 254

[2] Dereniak E L and Boreman G D 1996 *Infrared Detectors and Systems* vol 306 (New York: Wiley)

IOP Publishing

Principles of Biophotonics, Volume 2
Light emission, detection, and statistics
Gabriel Popescu

Chapter 3

Photon-based radiometric quantities

3.1 Number of photons

The radiometric quantities discussed so far are *energy-based*, that is, they describe the energy distribution in time, space, temporal frequency, spatial frequency, carried by an optical field or emitted by a source. An entirely equivalent set of radiometric quantities can be defined if we start with the number of photons, rather than energy [1, 2]. This description is valuable whenever the discrete nature of light is relevant, for example when studying photodetection. This equivalence stems from the fact that each photon carries an energy $h\nu$ or hc/λ, such that the total energy of optical radiation is (recall equation (2.1))

$$
\begin{aligned}
Q &= Nh\nu \\
N &= \frac{Q}{h\nu}.
\end{aligned}
\tag{3.1}
$$

In equation (3.1), N denotes the number of photons (unitless).

3.2 Photon density

The photon density is defined as the number of photons per unit volume,

$$
\begin{aligned}
n &= \frac{N}{V} \\
[n] &= \mathrm{m}^{-3}.
\end{aligned}
\tag{3.2}
$$

3.3 Photon flux

The photon quantity analogous to power is the photon flux,

doi:10.1088/978-0-7503-1644-6ch3

$$P_q = \frac{dN}{dt}$$
$$[P_q] = s^{-1}. \tag{3.3}$$

We use the same symbol for photon flux as for power, but the subscript q indicates a *photon* quantity.

Note the connection between P_q and P (power) is straightforward,

$$P = \frac{dQ}{dt} = \frac{d(N \cdot h\nu)}{dt} = h\nu\frac{dN}{dt} = h\nu P_q. \tag{3.4}$$

3.4 Photon temporal power spectrum

The photon flux per unit temporal frequency describes a photon-based power spectrum,

$$S_q(\nu) = \frac{dP_q(\nu)}{d\nu}$$
$$[S_q(\nu)] = s^{-1}/\text{Hz}. \tag{3.5}$$

$S_q(\nu)$ describes how many photons per second there are in the infinitesimal interval $(\nu, \nu + d\nu)$.

Note that this quantity can be quite different in shape from the energy-based power spectrum, $S(\nu)$,

$$S_q(\nu) = \frac{d}{d\nu}\left[\frac{P(\nu)}{h\nu}\right]$$
$$= \frac{1}{h\nu}\left[S(\nu) - \frac{P(\nu)}{\nu}\right]. \tag{3.6}$$

As in the case of an energy-based power spectrum, we can define $S_q(\omega)$ and $S_q(\lambda)$, with respect to the angular frequency, $\omega = 2\pi\nu$, and wavelength $\lambda = c/\nu$, respectively,

$$S_q(\omega) = \frac{1}{2\pi}S_q(\nu)$$
$$S_q(\lambda) = -\frac{c}{\lambda^2}S_q(\lambda). \tag{3.7}$$

3.5 Photon intensity

Photon intensity is the photon flux per unit solid angle, Ω, namely,

$$\begin{aligned} I_\Omega^q &= \frac{dP_q(\Omega)}{d\Omega} \\ \left[I_\Omega^q\right] &= \mathrm{s}^{-1}/\mathrm{srad}. \end{aligned} \tag{3.8}$$

As discussed in section 2.5, the solid angle relates to the wavevector,

$$\begin{aligned} d\Omega &= \frac{\hat{\mathbf{n}} \cdot d\mathbf{A}}{r^2} \\ &= \frac{dk_x dk_y}{k^2}. \end{aligned} \tag{3.9}$$

As a result, we can write the photon intensity as a function of the spatial power spectrum of the photon flux,

$$\begin{aligned} I_\Omega^q &= k^2 \frac{d^2 P_q(\mathbf{k}_\perp)}{d\mathbf{k}_\perp^2} \\ &= k^2 S_q(\mathbf{k}_\perp). \end{aligned} \tag{3.10}$$

In equation (3.10), $S_q(\mathbf{k}_\perp)$ is the spatial power spectrum of the photon flux, meaning the photon flux per infinitesimal wavevectors interval, $d^2\mathbf{k}_\perp = dk_x dk_y$.

3.6 Photon irradiance

Photon irradiance defines the photon flux delivered per unit areas

$$\begin{aligned} I_q &= \frac{dP_q}{dA} \\ [I_q] &= \frac{\mathrm{s}^{-1}}{\mathrm{m}^2}. \end{aligned} \tag{3.11}$$

Energy-based irradiance relates to I_q as

$$I = \frac{dP}{dA} = h\nu \frac{dP_q}{dA} = h\nu I_q. \tag{3.12}$$

3.7 Photon spectral irradiance

The photon spectral irradiance defines the photon flux per unit area, per unit of optical frequency.

$$\begin{aligned} I_\nu^q &= \frac{d^2 P_q}{dA d\nu} \\ \left[I_\nu^q\right] &= \frac{\mathrm{s}^{-1}}{\mathrm{m}^2\ \mathrm{Hz}}. \end{aligned} \tag{3.13}$$

3.8 Photon radiance

The photon radiance describes the photon flux emitted or absorbed by a surface, per unit solid angle, per unit projected area,

$$l_q = \frac{d^2 P_q(\Omega, A)}{d\Omega dA \cos\theta}$$
$$[l_q] = \text{s}^{-1}/\text{srad} \cdot \text{m}^2. \tag{3.14}$$

As before (section 2.8), $dA \cos\theta$ is the projection of the area normal to the direction of interest.

3.9 Photon spectral radiance

The photon spectral radiance describes the photon radiance per unit frequency,

$$l_v^q = \frac{dl_q}{dv}$$
$$\left[l_v^g\right] = \frac{\text{s}^{-1}}{\text{m}^2\ \text{srad Hz}}. \tag{3.15}$$

3.10 Photon exitance

The photon exitance is defined as the amount of photon flux emitted by a source per unit area,

$$M_q = \frac{dP_q}{dA_s}$$
$$[M_q] = \frac{\text{s}^{-1}}{\text{m}^2}. \tag{3.16}$$

Photon exitance describes a source and should be distinguished from the photon irradiance, which is a property of the field and shares the same units.

3.11 Photon spectral exitance

The photon spectral exitance is the photon exitance per unit frequency,

$$M_v^q = \frac{dM_2}{dv}$$
$$\left[M_v^q\right] = \frac{\text{s}^{-1}}{\text{m}^2\ \text{Hz}}. \tag{3.17}$$

3.12 Problems

1. Does the photon radiance satisfy the same conservation theorem as the (energy-based) radiance?
2. A photon has a wavelength $\lambda = 1$ µm. Calculate
   a) photon energy (J)
   b) photon energy in eV
   c) photon frequency (Hz).
   d) How many photons per second contribute to make 1 mW of power?

References

[1] Dereniak E L and Boreman G D 1996 *Infrared Detectors and Systems* vol 306 (New York: Wiley)

[2] Boyd R W 1983 *Radiometry and the Detection of Optical Radiation* (Wiley Series in Pure and Applied Optics) (New York: Wiley), vii p 254

IOP Publishing

Principles of Biophotonics, Volume 2
Light emission, detection, and statistics
Gabriel Popescu

Chapter 4

Photometric properties of light

Photometry deals with the properties of the light as perceived by the human eye as a *subjective* photodetector (for a fuller description of the subject, see [1] and volume 2 in [2]). In contrast to radiometry, photometry takes into account the spectral sensitivity of the eye, in other words, the fact that the eye has a different sensitivity for different colors. The eye spectral response is called the *luminosity function* and for the normal eye, has two distinct forms, for daytime (*photopic*) and dark time (*scotopic*) adaptation (see figure 4.1). Note that the eye's response shifts to shorter wavelengths in the dark.

The daytime spectral response is used to define a set of quantities that is analogous to the radiometric ones, as follows (we will use the same familiar symbols, with subscript l, for 'luminous').

4.1 Luminous energy

Luminous energy defines the total energy weighted according to the *luminosity function* shown in figure 4.1. Luminous energy is measured in lumen second,

$$[Q_l] = \text{lm s}. \tag{4.1.}$$

Lumen is the unit for *luminous flux*, as defined next.

4.2 Luminous flux

The luminous flux (or luminous power) is the photometric equivalent of power. Like all photometric quantities, luminous flux is also weighted by the luminosity function.

$$\begin{gathered} P_l = \frac{dQ_l}{dt}. \\ [P_l] = \text{lm} \end{gathered} \tag{4.2}$$

Thus, *lumen* is the unit for luminous flux.

doi:10.1088/978-0-7503-1644-6ch4

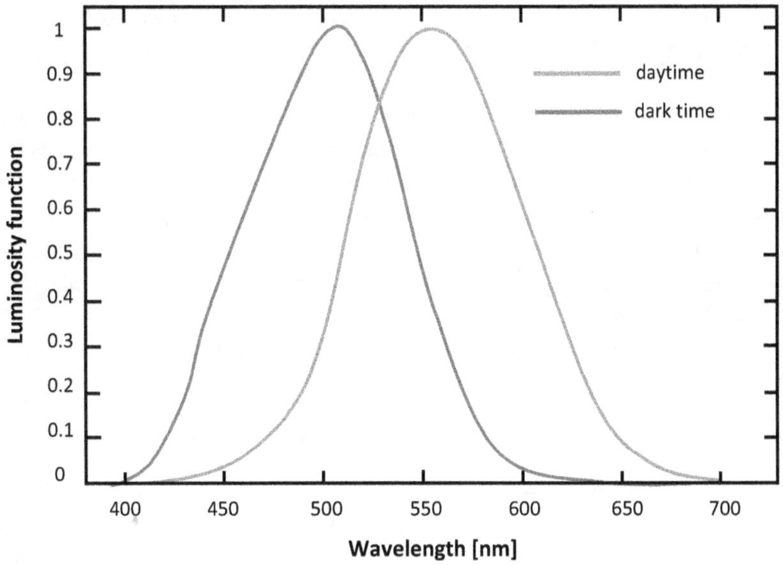

Figure 4.1. Photopic (daytime-adapted, green curve) and scotopic (darkness-adapted, blue curve) luminosity functions.

4.3 Luminous energy density

The luminous energy density describes the luminous energy per unit volume,

$$W_l = \frac{dQ_l}{dV}$$
$$[W_l] = \frac{\text{lm s}}{\text{m}^3}. \tag{4.3}$$

4.4 Luminous intensity

The *luminous intensity* represents the luminous flux per unit solid angle,

$$I_l^{\Omega} = \frac{dP_l}{d\Omega}$$
$$[I_l^{\Omega}] = \frac{\text{lm}}{\text{srad}}$$
$$= \text{cd (candela)}. \tag{4.4}$$

Note that *candela* (cd) is one of the seven primary units comprising the International System of Units (SI). The other six units are meter (m) for length, second (s) for time, kilogram (kg) for mass, ampere (A) for electric current, kelvin (K) for temperature, and mole (mol) for amount of substance.

Candela is defined with respect to (radiometric) intensity (see section 2.5), which can be measured using a photodetector. According to the current standard (the National Institute of Standards and Technology (NIST) reference on constants, units and uncertainty [3]): 'the candela is the luminous intensity in a given direction, of a source that emits monochromatic radiation of frequency 540×10^{12} Hz and that has a radiant intensity in that direction of 1/683 W srad^{-1}.

The wavelength corresponding to the frequency of 540×10^{12} Hz is $\lambda = c/\nu = 555$ nm and corresponds precisely to the maximum of the photopic luminosity function (figure 4.1). The human eye is the most sensitive at this wavelength, when adjusted for bright conditions.

The luminous intensity at any wavelength can be obtained as a function of the radiometric intensity as

$$I_l^{\Omega}(\lambda) = [I_{\Omega}(\lambda)\,Y(\lambda)]683\ \text{lm/W}, \tag{4.5}$$

where I_{Ω} is the radiometric intensity (in W/srad), and $Y(\lambda)$ is the photopic luminosity function evaluated at the particular wavelength λ.

4.5 Illuminance

Illuminance is the photometric equivalent to irradiance, that is, the luminous flux per unit area,

$$\begin{aligned} I_l &= \frac{dP_l}{dA} \\ [I_l] &= \frac{\text{lm}}{\text{m}^2} \\ &= lux,\ \ lx. \end{aligned} \tag{4.6}$$

The unit for illuminance is *lux*.

4.6 Luminance

Luminance is the equivalent of the photometric radiance. It measures the luminous power per solid angle, per normal unit area of the source:

$$\begin{aligned} L_l &= \frac{d^2P_l}{d\Omega\, dA\cos(\theta)} \\ [L_l] &= \frac{\text{lm}}{\text{srad m}^2} \\ &= \frac{\text{cd}}{\text{m}^2}. \end{aligned} \tag{4.7}$$

Brightness is sometimes used interchangeably with luminance.

4.7 Problems

1. Prove that in a lossless system, the brightness (luminance) is conserved.
2. An isotropic source emits 1 lm of luminous flux at wavelength $\lambda = 550$ nm.
   a) What is the illuminance at a distance of $L = 10$ m?
   b) If the human eye pupil is 2 mm in diameter, what is the flux reaching the retina?
   c) What is the new luminous flux detected by the retina, if $\lambda = 650$ nm?

References

[1] Walsh J W T 1958 Photometry *Photoniques* (Edinburgh: Constable)

[2] Bass M, Mahajan V N and Optical Society of America 2010 *Handbook of Optics* 3rd edn (New York: McGraw-Hill) v.<1, 4–5>

[3] NIST *Reference on constants, Units and Uncertainty* https://physics.nist.gov/cuu/Units/index.html

IOP Publishing

Principles of Biophotonics, Volume 2
Light emission, detection, and statistics

Gabriel Popescu

Chapter 5

Fluorescence

Luminescence, comprising fluorescence and phosphorescence, is the phenomenon by which a given substance emits light. Light in the visible spectrum is due to *electronic transitions. Vibrationally excited states* can give rise to lower energy photons, i.e. *infrared* radiation, and *rotational modes* produce even lower energy radiation, in the *microwave* region.

In recent years, tremendous progress has been made toward the use of fluorescence in biophysics and biomedicine. Fluorescence-based measurements are currently used in microscopy, flow cytometry, medical diagnosis, DNA sequencing, and many other areas [1–5]. Most importantly for biophotonics, fluorescence microscopy is a powerful tool for imaging at the cellular and molecular scale [1]. Fluorescent probes can specifically bind to molecules of interest, revealing not only localization information, but also activity (i.e., *function*). Recent developments have pushed fluorescence microscopy towards the single molecule scale. A grand challenge in microscopy is to merge information from the molecular and cellular scales.

5.1 Jablonski diagram

Fluorescence is defined as the radiative transition between two electronic states of the same *spin multiplicity*. The spin multiplicity is defined as $M_s = 2s + 1$, with s the spin of the state. Since each electron can have a spin of either 1/2 or −1/2 (because electrons are *fermions)*, the spin, s, is a number that describes the sum of the spins from all the electrons on that electronic state. Thus, $s = 0$ when there are two paired electrons on the electronic sate, meaning the two electrons have antiparallel spins. The spin takes the value $s = 1/2$ when there is a single, unpaired electron in the state. Finally, $s = 1$ when there are two electrons of parallel spins. Therefore, electronic states can be *singlets* ($M_s = 2s + 1 = 1$), *doublets* ($M_s = 2$), or *triplets* ($M_s = 3$). Most organic molecules have paired electrons in their ground state: in *singlet* states, the pairing electrons have their spins *antiparallel*, $s = 1/2 - 1/2 = 0$, $2s + 1 = 1$, and in

doi:10.1088/978-0-7503-1644-6ch5

triplet states they are parallel, $s = 1/2 + 1/2 = 1$, $2s + 1 = 3$. The *doublet* represents a state with an unpaired electron, $s = 1/2$, $2s + 1 = 2$. The singlet–singlet transition is the most common in fluorescence.

The phenomenon of fluorescence is commonly described via a *Jablonski diagram*, in which the energy levels of a molecule are drawn as horizontal lines ordered along the vertical axis according to their energy (figure 5.1). This method of illustrating the energy levels is reminiscent of a musical score. Perhaps this is not surprising, as Jablonski himself was an excellent violinist: between 1921 and 1926 he played for the Warsaw Opera as the first violin (while pursuing his doctorate degree!) [4]. In figure 5.1, S_0, S_1, and S_2 denote the ground, first, and second electronic state, respectively, all assumed to be singlets. Each electronic state contains a number of *vibrational levels*, denoted by 0, 1, 2, etc. The vertical lines indicate transitions between various states. A typical scenario is as follows: the molecule in ground state S_0 absorbs one photon of energy $h\omega_A/2\pi = h\nu_A$, and occupies an excited state S_2 (with Planck's constant, $h = 6.6 \times 10^{-34}$ Js). Due to *internal conversion*, the molecule decays on the lower vibrational level S_1, within 1 ps $= 10^{-12}$ s or less. After an average time of 10^{-9}–10^{-8} s, the molecule *radiatively decays* to the ground state, with emission of a photon, of energy $h\nu_F$. Thus, since this fluorescence lifetime of 10^{-8} s is much larger than 1 ps, we can assume that the internal conversion is complete prior to emission.

Absorption occurs at the 1 fs $= 10^{-15}$ s scale, which is too fast for significant displacement of the nuclei during the vibrational motion. This assumption, referred to as the *Born–Oppenheimer approximation*, simplifies the calculations of a molecule's wave function, and allows it to factorize into electronic and nuclear (*vibrational* and *rotational*) components

$$\psi_{\text{total}} = \psi_{\bar{e}} \times \psi_N. \tag{5.1}$$

Molecules in the S_1 state can also undergo a spin conversion to the first triplet state T_1. Emission from T_1 is termed *phosphorescence* and is generally shifted to lower energy (longer wavelengths). The conversion from S_1 to T_1 is referred to as *intersystem crossing*. Transition $T_1 \rightarrow S_0$ is forbidden, thus the rate constants for triplet emission, called *phosphorescence*, are orders of magnitude smaller than those

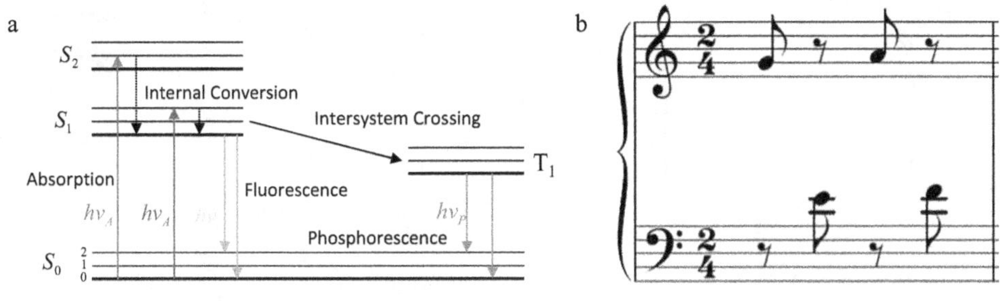

Figure 5.1. (a) Jablonski diagram depicting a three-level system: S_0, S_1, S_2 are singlet electronic states, each containing vibrational levels. Absorption takes place at frequencies υ_A. T_1 denotes a triplet state, from which phosphorescence takes place. (b) Musical score illustrating the similarity with the Jablonski diagram.

for fluorescence. It is common to observe phosphorescence with decay times of seconds or even minutes. Molecules of heavy atoms tend to be phosphorescent.

5.2 Emission spectra

There exists a symmetry property (sometimes called 'mirror symmetry') between the power spectrum of absorption (or excitation) and emission (or radiative decay), as follows. In the excitation sequence, absorption from S_0 is followed by transition to higher vibrational levels of, say, S_1, and finally *nonradiative* decay to the lowest vibrational level within S_1. During emission, the molecule *radiatively* decays from S_1 to a high vibrational level within S_0, followed by thermal equilibration to the ground vibrational state. This process results in a symmetric shape of absorption and emission spectra (figure 5.2).

Note that the local maxima are due to the individual vibration levels. A general property of fluorescence is that the emission spectrum remains constant irrespective of the excitation wavelength (exceptions exist). This happens because, upon excitation to higher vibrational levels of S_1, the excess energy is quickly dissipated and the molecule drops to the lowest vibrational level. Emission occurs predominantly from the lowest singlet state (S_1, in our example). Interestingly, the emission

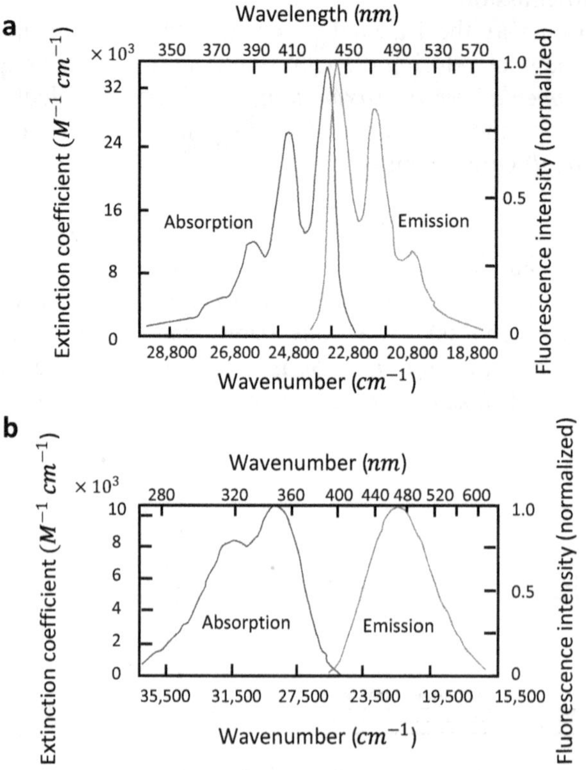

Figure 5.2. (a) Mirror image rule (perylene in benzene). (b) Exception to the mirror rule (quinine sulfate in H_2SO_4). Adapted from [4], copyright Springer Science & Business Media.

spectrum of quinine (figure 5.2(b)) mirrors only the larger wavenumber peak of the absorption spectrum, associated with the $S_0 - S_1$ transition. The first (shorter wavenumber) peak is due to the transition $S_0 - S_2$. Since the emission occurs predominantly from S_1, the second peak is absent, which apparently breaks the *mirror rule*. By contrast, in the case of perylene (figure 5.2(a)), the peaks in both absorption and emission correspond to vibrational levels, which results in perfect symmetry.

5.3 Rate equations

The population of each level (say, S_0, S_1, S_2 of the three-level system illustrated in figure 5.3) is determined by the rate of *absorption*, *nonradiative decay*, *spontaneous emission*, and *stimulated emission*. This description is due to Einstein, who recognized the need for the stimulated emission term, which is the radiative transition from an excited state to the ground state *triggered* (stimulated) by the excitation field. The concept of stimulated emission, by which one excitation photon results in two identical emitted photons, established the groundwork for the development of lasers (see chapter 7). For the two levels, S_0 and S_1, we can write the rate equations as:

$$\frac{dN_1}{dt} = B_{01}N_0 - B_{10}N_1 - A_{10}N_1 - \gamma_{nr}N_1 \tag{5.2a}$$

$$\begin{aligned}\frac{dN_0}{dt} &= -B_{01}N_0 + B_{10}N_1 + A_{10}N_1 + \gamma_{nr}N_1 \\ &= -\frac{dN_1}{dt}.\end{aligned} \tag{5.2b}$$

In equations (5.2*a* and *b*), the concentrations $N_{0,1}$ are in units of m^{-3}. All the rate constants, in units of s^{-1}, are defined as follows: B_{01} is the absorption rate, B_{10} is the stimulated emission rate ($B_{01} = B_{10}$ for nondegenerate levels), A_{10} is the spontaneous emission rate, and γ_{nr} is the rate of nonradiative decay. Note that the energy lost through all the possible channels of nonradiative decay, generically captured by γ_{nr}, eventually converts into heat. At thermal equilibrium, $N_1 \ll N_0$, and, as a result, the stimulated emission term can be neglected ($B_{10}N_1 \simeq 0$). This is not the case, of course, when dealing with an *active medium*, where $N_1 > N_0$, that is, when there is *population inversion* (see chapter 7). Furthermore, in equation (5.2b), we neglected the probability of having molecules in other states, that is, we assumed

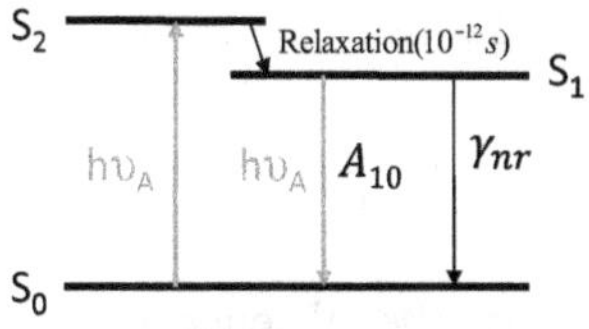

Figure 5.3. Jablonski diagram for a three-level system.

$N_0 + N_1 = \text{const.}$, or $d(N_0 + N_1)/dt = 0$. Thus, if we assume that the total number of molecules is known, $N_0 + N_1 = N$, we can easily solve the system of equations at *steady state*, when $\frac{dN_1}{dt} = \frac{dN_2}{dt} = 0$, namely,

$$\begin{aligned} B_{01}N_0 - A_{10}N_1 - \gamma_{nr}N_1 = 0 \\ N_0 + N_1 = N. \end{aligned} \tag{5.3}$$

The solutions for the populations of the two levels can be readily obtained as

$$\begin{aligned} N_0 &= N\frac{A_{10} + \gamma_{nr}}{B_{01} + A_{10} + \gamma_{nr}} \\ N_1 &= N\frac{B_{01}}{B_{01} + A_{10} + \gamma_{nr}}. \end{aligned} \tag{5.4}$$

Note that, at thermal equilibrium, $N_0 \gg N_1$, that is, the rate of decay is much higher than that of absorption, $A_{01} + \gamma_{nr} \gg B_{01}$.

5.4 Quantum yield

The *fluorescence quantum yield* is the ratio of the number of photons emitted to that of photons absorbed,

$$\begin{aligned} Q &= \frac{N_1A_{10}}{N_0B_{01}} \\ &= \frac{A_{10}}{A_{10} + \gamma_{nr}}, \end{aligned} \tag{5.5}$$

where we use that, at thermal equilibrium, equations (5.4) yield $N_1/N_0 = B_{01}/(A_{10} + \gamma_{nr})$. Thus, the quantum yield can approach unity if the non-radiative decay, γ_{nr}, is much smaller than the rate of radiative decay. Note that even for 100% quantum yield, that is, where the number of emitted photons equals that of absorbed photons, the energy conversion is always less than unity because the wavelength of the fluorescent light is longer, or, equivalently, the frequency is lower, that is $h\omega_A > h\omega_F$.

5.5 Fluorescence lifetime

The *lifetime* of the excited state is defined by the average time the molecule spends in the excited state before returning to the ground state. For a simplified version of the Jablonski diagram illustrated in figure 5.3, the lifetime is defined as the inverse of the decay rate, namely

$$\tau = \frac{1}{A_{10} + \gamma_{nr}}. \tag{5.6}$$

The lifetime of a fluorophore in the absence of nonradiative processes, that is, $\gamma_{nr} \to 0$, is called the *natural lifetime*,

$$\tau_0 = \frac{1}{A_{10}}. \tag{5.7}$$

The natural lifetime, τ_0, depends on the absorption spectrum, extinction coefficient, and emission spectrum of the fluorophore. The measured lifetime, τ, always incorporates the effects of nonradiative processes, which increase the probability of decay and, thus, is shorter than the natural lifetime. The measured and natural lifetimes are related via the quantum yield, Q,

$$\begin{aligned} \tau &= \tau_0 \frac{A_{01}}{A_{01} + \gamma_{nr}} \\ &= Q\tau_0 \\ \tau_0 &= \frac{\tau}{Q}, \end{aligned} \tag{5.8}$$

where Q is defined by equation (5.5).

5.6 Quenching

Fluorescence quenching represents the decay to the ground state via different mechanisms that do not involve radiation: in other words, all processes that result in nonradiative decay. *Collisional* quenching is due to the contact between the fluorophore and some other molecule in the solution, called the *quencher*. During this collision, which is the result of thermal diffusion, the molecules are assumed to remain intact. Examples of quenchers include: halogens, oxygen, amines, etc. Aside from collisions, there are other processes that can generate quenching, for example, the fluorophore can form a *nonfluorescent* complex with the quencher (*static quenching*). Another common quenching mechanism is due to the attenuation of the incident light by the fluorophore itself, or another absorbing species.

The photon absorption takes place at the femtosecond scale (10^{-15} s). Thus, *absorption spectroscopy* does not report on molecular dynamics, but rather on the average ground state of the molecules that absorb light. By contrast, emission occurs over a larger period of time. Therefore, during the time that fluorescent molecules remain in the excited state, interaction with various quenchers can report on their diffusion properties. For example, oxygen in water has a diffusion coefficient of $D = 2.5 \cdot 10^{-5}\ \mathrm{m^2/s}$. If a fluorophore has a lifetime of 10 ns, the root mean squared distance that the oxygen molecule can travel during this time is given by Einstein's equation

$$\langle \Delta x^2 \rangle = 2Dt. \tag{5.9}$$

Thus, $\sqrt{\langle \Delta x^2 \rangle} = 7$ nm, which is a significant distance, comparable with the thickness of a cell membrane bilayer. Since the cell lipid bilayer is only 4–5 nm thick, this estimation also reveals that it is possible for a molecule to absorb light outside the cell and fluoresce from the inside (and vice versa), one lifetime later. This estimation indicates that, during its lifetime, a fluorophore can interact with quenchers from within a radius of 7 nm.

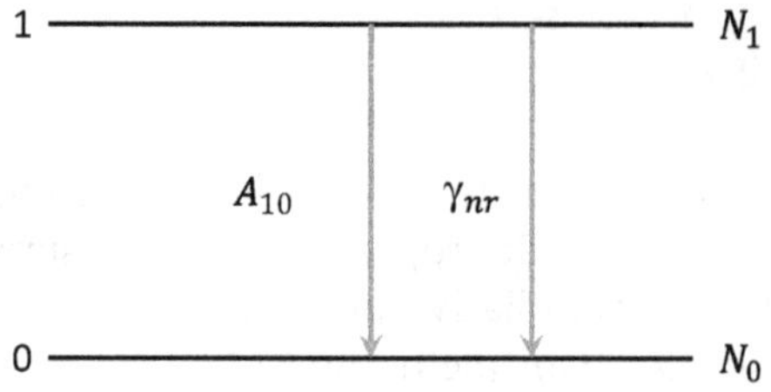

Figure 5.4. Problem 5.1.

5.7 Problems

1. Consider the two-level system shown in figure 5.4. The population of the excited state decays as

$$\frac{dN_1(t)}{dt} = -(A_{10} + \gamma_{nr})N_1(t)$$

where A_{10} is the spontaneous emission rate and γ_{nr} is the nonradiative decay. Solve for $N_1(t)$ and find the lifetime of the upper state.

2. The intensity associated with fluorescence decay for a certain transition follows the decay

$$I(t) = I_0 e^{-\frac{t}{\tau}}, \quad t > 0.$$

Prove that the *average time*, $<t>$, of the intensity distribution, equals the decay constant, τ, which is the lifetime of the excited state,

$$<t> = \tau.$$

3. For a particular fluorescent process, the intensity decay can be approximated as a sum of two exponentials,

$$I(t) = I_1 e^{-\frac{t}{\tau_1}} + I_2 e^{-\frac{t}{\tau_2}}, \quad t > 0.$$

Compute the mean time, $<t>$, for this decay in terms of τ_1, τ_2.

4. Generalize the result in problem 3 for an intensity decay of the form

$$I(t) = \sum_{n=1}^{N} I_n e^{-\frac{t}{\tau_n}}, \quad t > 0.$$

5. The fluorescence decay excited by an infinitely short pulse has the form

$$I(t) = I_0 e^{-\frac{t}{\tau_0}}.$$

Calculate the intensity decay if the excitation pulse has the form

$$p(t) = p_0 e^{-\frac{t}{\tau_p}}.$$

If $I_0 = 1, p_0 = 1, \tau = 5$ ns, plot the intensity decay for $\tau_p = 1$ ps, 1 ns, 10 ns, 1 μs.

References

[1] Diaspro A 2011 *Optical Fluorescence Microscopy: From the Spectral to the Nano Dimension* (Berlin: Springer)

[2] Digman M A and Gratton E 2009 Fluorescence correlation spectroscopy and fluorescence cross-correlation spectroscopy *WIREs Syst. Biol. Med.* **1** 273–82

[3] Geddes C D 2011 *Reviews in Fluorescence* (New York: Springer)

[4] Lakowicz J R 2006 *Principles of Fluorescence Spectroscopy* (New York: Springer), xxvi p 954

[5] Mycek M-A and Pogue B W 2003 *Handbook of Biomedical Fluorescence* (New York: Marcel Dekker), xv p 665

IOP Publishing

Principles of Biophotonics, Volume 2
Light emission, detection, and statistics

Gabriel Popescu

Chapter 6

Black body radiation

The most commonly encountered electromagnetic radiation is of thermal origin. A *black body* is an idealized model of a physical object that absorbs *all incident* electromagnetic radiation. Because it is a perfect *absorber* at all wavelengths, a black body is also an ideal *emitter* of thermal radiation. This *black body radiation* has a certain frequency (or wavelength) distribution, which is characterized by a maximum. This spectral distribution of radiation by bodies at thermal equilibrium was a problem of critical interest at the turn of the 20th century, and led to the development of quantum mechanics.

Thermal radiation is emitted by a body that exists at a temperature higher than absolute zero. In essence, this radiation is generated by converting the internal energy of the body at thermal equilibrium and represents the reverse process to absorption. The spectral content of the radiation is determined by the *mode distribution*, that is, the spatial frequency content of the electromagnetic field within a certain bounded space, or *cavity* (figure 6.1). A mode of the electromagnetic field in the cavity satisfies the condition of vanishing electric field at the wall. Clearly, as the wavelength decreases, there are increasingly more ways of 'fitting' the modes in the cavity. The formula that correctly predicts the thermal radiation by a black body was derived by Planck in 1900 [1], as described below.

6.1 Planck's radiation formula

Let us consider a radiating cavity, with its dimensions much larger than the wavelength of light (see, e.g., [2]). The problem of finding the spectral distribution, $du(\nu)/d\nu$ (energy per unit frequency, ν), of the radiation emitted by this cavity, approximated by a black body, comes down to calculating the number of modes, $d\nu$, into the volume V, that exist within a certain frequency range $d\nu$. The radiated energy per mode, per unit volume is

doi:10.1088/978-0-7503-1644-6ch6

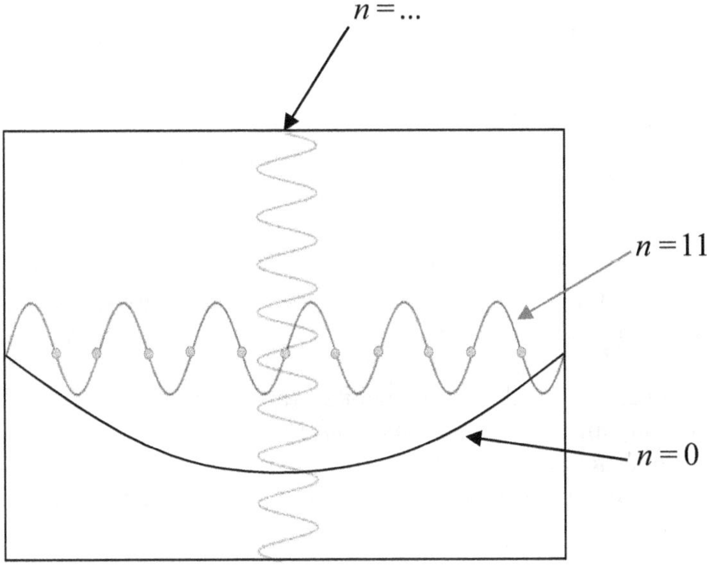

Figure 6.1. Cavity modes: the sinusoids represent the real part of the electric field. All surviving field modes have zero values at the boundary. The modes are indexed by n, the number of zeros along an axis (orange dots).

$$du(\nu) = 2f\frac{h\nu}{V}dN. \tag{6.1}$$

In equation (6.1), f is the *probability of occupancy* associated with a given mode, h is Planck's constant ($h = 6.6 \cdot 10^{-34}$ J s), and the factor two accounts for the two polarization modes that can exist in the cavity. The expression for f is obtained from the *Bose–Einstein statistics* that apply to indistinguishable particles with an unlimited state of occupancy, i.e. not obeying Pauli's exclusion principle [2],

$$f = \frac{1}{e^{\frac{h\nu}{k_B T}} - 1}, \tag{6.2}$$

where k_B is Boltzmann's constant, $k_B = 1.38 \cdot 10^{-23}$J/K. The average energy per mode is

$$\begin{aligned}\langle E\rangle &= fh\nu \\ &= \frac{h\nu}{e^{\frac{h\nu}{k_B T}} - 1}.\end{aligned} \tag{6.3}$$

Planck's formula predicts the spectral density of the radiation emitted by a black body, at thermal equilibrium, as a function of temperature. Let us assume a cavity of size much larger than the wavelength of light. Black body radiation applies to an object that absorbs all radiation incident to it, and re-radiates energy that depends

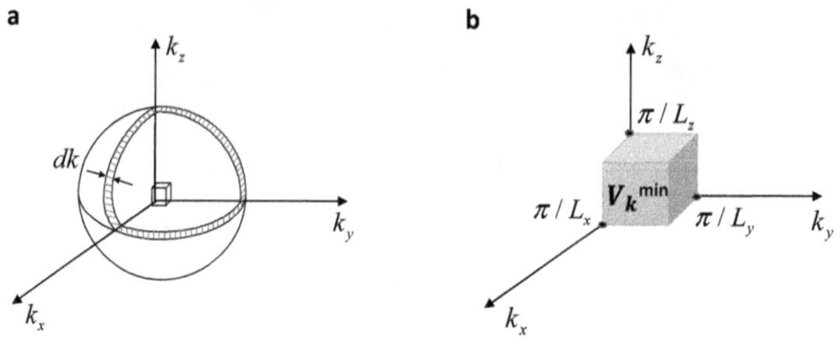

Figure 6.2. a) Mode distribution in a cavity. b) The cube at the origin in a) is the smallest volume in **k**-space, defined by the inverse dimensions of the cavity.

only on its temperature and not the incident radiation. The radiated field can be considered as consisting of the *resonant modes* of the cavity.

In order to find the number of modes per frequency interval, let us consider the wavevector space in figure 6.2(a). The spherical shell in the first octant of the **k**-space is

$$\begin{aligned} dV_k &= \frac{1}{8} 4\pi k^2 \, dk \\ &= \frac{\pi}{2} k^2 \, dk. \end{aligned} \tag{6.4}$$

The number of modes within this interval is

$$dN = \frac{dV_k}{V_k^{\min}}, \tag{6.5}$$

where $V_k^{\min}$ is the volume in **k**-space formed by the smallest spatial frequencies, $V_k^{\min} = \frac{\pi^3}{L_x L_y L_z}$, with $L_x L_y L_z = V$ the volume of the cavity (figure 6.2(b)). Equation (6.5) can now be expressed as

$$dN = \frac{\pi}{2} \frac{V}{\pi^3} k^2 \, dk. \tag{6.6}$$

In order to find the number of modes dN per frequency interval, $d\nu$, rather than dk, we use the *dispersion relation*, namely, that the magnitude of the wavevector, $k = |\mathbf{k}|$ equals the wave number in vacuum,

$$k = 2\pi \frac{\nu}{c}, \tag{6.7}$$

Combining equations (6.6) and (6.7), we obtain

$$\begin{aligned} dN &= \frac{\pi}{2}\frac{V}{\pi^3}\frac{4\pi^2\nu^2}{c^2}\frac{2\pi}{c}d\nu \\ &= \frac{4\pi V}{c^3}\nu^2\, d\nu. \end{aligned} \tag{6.8}$$

Finally, plugging equation (6.8) into equation (6.1), we obtain the *radiated energy per unit frequency range and unit volume*, $\rho(\nu)$, that is, Planck's formula derived in 1900 (for a review of Planck's work on the theory of heat radiation, see [1]):

$$\begin{aligned} \rho(\nu) &= \frac{du(\nu)}{d\nu} \\ &= \frac{8\pi h}{c^3}\frac{\nu^3}{e^{\frac{h\nu}{k_B T}} - 1}. \end{aligned} \tag{6.9}$$

Equation (6.9) is fundamental to calculating any quantity related to black body radiation. For example, from equation (6.9), we can calculate the power flowing from the cavity, through a surface A and element of solid angle $d\Omega$, per unit of frequency (see figure 6.3),

$$\begin{aligned} d^2P(\nu) &= \frac{c}{2}du(\nu)\frac{A\cos\theta\, d\Omega}{2\pi} \\ &= \frac{c}{2}\rho(\nu)d\nu A\cos\theta\frac{d\Omega}{2\pi} \end{aligned} \tag{6.10}$$

where A is the area, θ is the angle with respect to the surface normal, and the solid angle element, $d\Omega = 2\pi\sin\theta\, d\theta$. Integrating on the hemisphere, we obtain the total power

$$\begin{aligned} dP(\nu) &= \frac{c}{2}du(\nu)A\int_0^{\pi}\cos\theta\sin\theta\, d\theta \\ &= \frac{1}{4}cA\, du(\nu) \\ &= \frac{1}{4}cA\rho(\nu)d\nu. \end{aligned} \tag{6.11}$$

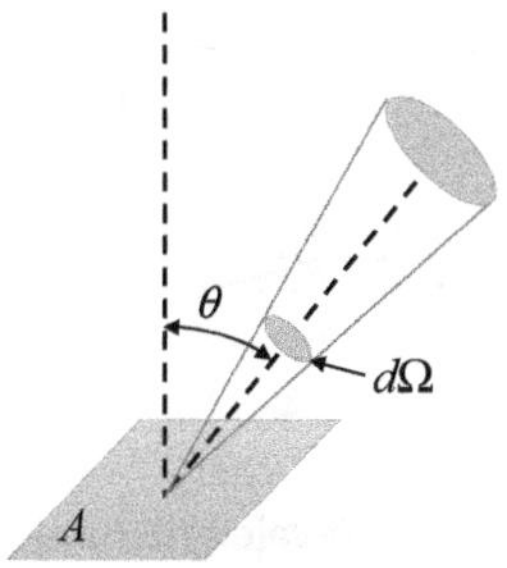

Figure 6.3. Power flow out of a black body surface.

Further, the *spectral exitance*, that is, the power per surface area per frequency (see section 2.11), $M_\nu = \frac{d^2P}{dA\, d\nu}$, has the form

$$\begin{aligned} M_\nu &= \frac{1}{4}c\rho(\nu) \\ &= \frac{2\pi h}{c^2}\frac{\nu^3}{e^{\frac{h\nu}{kT}} - 1}. \end{aligned} \tag{6.12}$$

The spectral exitance can be expressed in terms of the wavelength as

$$\begin{aligned} M_\lambda &= M_\nu(\nu = c/\lambda)\frac{d\nu}{d\lambda} \\ &= \frac{2\pi hc^2}{\lambda^5}\frac{1}{e^{\frac{hc}{\lambda kT}} - 1}. \end{aligned} \tag{6.13}$$

Note that simply replacing ν with λ in equation (6.12) yields the wrong formula. The Jacobian factor, $\frac{d\nu}{d\lambda}$, is very important, as it ensures that the M_λ is a *distribution.* One way to remember the change from a frequency to a wavelength distribution is that the exitance in each infinitesimal range is constant, namely,

$$M_\lambda\, d\lambda = M_\nu\, d\nu. \tag{6.14}$$

Figure 6.4 illustrates M_λ for different temperatures.

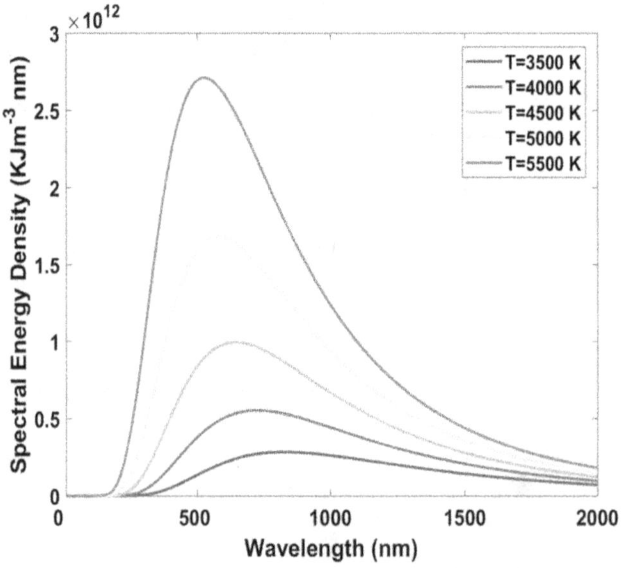

Figure 6.4. Spectral exitance of black bodies at different temperatures.

6.2 Wien's displacement law

We note immediately that both the energy per unit volume and unit frequency, $\rho(\nu)$, and the spectral exitance, M_ν, the power per surface area per frequency, exhibit a maximum at a particular frequency, $\nu_{\max}$, which is a function of temperature (see figure 6.5),

$$\left.\frac{d\rho(\nu)}{d\nu}\right|_{\nu=\nu_{\max}} = \left.\frac{dM_\nu}{d\nu}\right|_{\nu=\nu_{\max}} = 0. \tag{6.15}$$

The dependence of $\nu_{\max}$ on temperature is known as *Wien's displacement law* (see problem 6.2),

$$\nu_{\max} \propto T.$$

An equivalent way of expressing the displacement law is via the wavelength of the maximum emission,

$$\lambda_{\max} \propto \frac{a}{T},$$

where a is a constant, a = 2900 μm K. Note that $\lambda_{\max}$ is obtained by finding the maximum of the function M_λ with respect to λ (see problem 6.2). Simply substituting $\nu_{\max}$ yields the wrong result, in other words, $\lambda_{\max} \neq c/\nu_{\max}$.

This relationship between the temperature of the source and the peak wavelength of its emission led to some researchers expressing the 'color' of thermal light by the

Temperature	Source
1850 K	Candle flame, sunset/sunrise
2400 K	Standard incandescent lamps
2700 K	"Soft white" compact fluorescent and LED lamps
3000 K	"Warm white" compact fluorescent and LED lamps
3200 K	Studio lamps, photofloods, etc.
5000 K	Compact fluorescent lamps (CFL)
6200 K	Xenon short-arc lamp
6500 K	Daylight, overcast
6500 – 9500 K	LCD or CRT screen
15,000 – 27,000 K	Clear blue sky

Figure 6.5. Color temperature for various thermal sources.

temperature of the source. For example, *color temperature* is a typical measure on commercial microscopes when adjusting the power of the illuminating lamp. It has become customary to compare light sources (light bulbs, light emitting diodes, incandescent lamps, computer monitors, etc) in terms of their color temperatures. Thus, the color temperature defines which black body radiation would most closely match the light in question, from 'warm reddish' to 'cool blueish' (see figure 6.5). For example, the blue–white fluorescent light in most offices may have a color temperature of 5000 K, while an incandescent bulb has a temperature of 2000–2500 K, giving it a more reddish, 'warmer' appearance.

We experience Wien's displacement formula in our daily activities, as follows. (1) The Sun's effective temperature is 5800 K, which places its peak emission at ~500 nm (green), near the maximum sensitivity of our eye. This fact suggests that humans evolved to gain maximum sensitivity of their visual system at the most dominant wavelength emitted by the Sun. (2) Dimming the light on an incandescent light bulb will result in shifting the color toward red (longer wavelengths). (3) Heating a piece of metal will eventually produce radiation, first of red color and then blue–white when the temperature increases further. One can say that 'white-hot' is hotter than 'red-hot'. (4) Warm-blooded animals at, say, $T = 310$ K (37 °C), emit peak radiation at ~10 μm, in the infrared region of the spectrum, outside our eye sensitivity. Some reptiles and specialized cameras can sense these wavelengths and, thus, detect the presence of such animals. (5) Wood fire can have temperatures of 1500–2000 K, with peak radiation at ~2–2.5 μm. This means that most of the radiation is in the infrared spectrum, which we sense as heat, but only a small portion of the spectrum is visible.

6.3 Stefan–Boltzmann law

Another fundamental property of black body radiation is that the frequency-integrated spectral exitance, meaning, the *exitance* (in W/m^2), is proportional to the fourth power of temperature (proof left as an exercise, see problem 6.1),

$$\begin{aligned} M &= \int_0^\infty M_\nu \, d\nu \\ &= \sigma T^4. \end{aligned} \tag{6.16}$$

Equation (6.16) is known as the *Stefan–Boltzmann law*, and the proportionality constant, or the *Stefan–Boltzmann constant*, has the value $\sigma = 5.67 \cdot 10^{-8}$ Wm^{-2} K^{-4}.

The total power emitted by a black body is MA, where A is the area of the source. As a result, the Stefan–Boltzmann law becomes a practical means to estimate the size of other stars. Thus, the total power emitted by the Sun (S) and the star of interest (x), are, respectively

$$P_S = 4\pi R_S{}^2 \sigma T_S{}^4$$

and

$$P_x = 4\pi R_x{}^2 \sigma T_x{}^4.$$

The unknown radius can be easily obtained as

$$R_x = R_S\left(\frac{T_S}{T_x}\right)^2\sqrt{\frac{P_x}{P_S}}.$$

Note that the effective temperature of the star can be measured from the spectral distribution of the radiation fitted with Planck's formula.

A body that does not absorb all the incident radiation emits less total energy than a black body and is sometimes called a *gray* body. These bodies are characterized by an emissivity, $\varepsilon < 1$, such that the exitance is scaled down as

$$M = \varepsilon\sigma T^4.$$

6.4 Asymptotic behaviors of Planck's formula

Investigating Planck's formula we can readily find two asymptotic behaviors for the black body radiation, as follows. At *low temperatures*, $h\nu \gg k_BT$, we obtain

$$\begin{aligned} f &\simeq e^{-\frac{h\nu}{k_BT}} \\ \rho(\nu) &\simeq \frac{8\pi h}{c^3}\nu^3 e^{-\frac{h\nu}{k_BT}}. \end{aligned} \tag{6.17}$$

This formula approximates well the high frequency portion of the curve (figure 6.6). This behavior is known as the Wien approximation.

At *high temperatures*, $h\nu \ll k_BT$, the following approximations apply

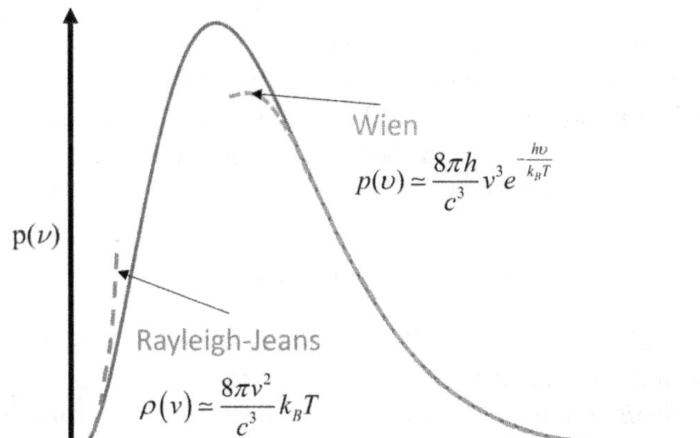

Figure 6.6. The asymptotic behavior of Planck's formula (blue curve) for high temperature (low-frequency, Rayleigh–Jeans, red curve) and low-temperature (high-frequency, Wien, green curve).

$$f \simeq \frac{1}{1 + \dfrac{h\nu}{k_B T} - 1} \tag{6.18a}$$

$$= \frac{k_B T}{h\nu}$$
$$\rho(\nu) \simeq \frac{8\pi\nu^2}{c^3} k_B T. \tag{6.18b}$$

Equation (6.18*a* and *b*) is known as the *Rayleigh–Jeans law* and describes well the low-frequency curve of Planck's equation (figure 6.6). Note that the Rayleigh–Jeans law is known as the *classic limit* formula. It strongly disagrees with Planck's law at high frequencies: one consequence of the Rayleigh–Jeans formula is that the amount of energy radiated over the entire spectral range diverges. This was known as the 'ultraviolet catastrophe' and was the main motivation behind developing a better understanding of black body radiation. Planck's work elucidated the problem and, at the same time, opened the door for quantum physics.

6.5 Einstein's derivation of Planck's formula

Einstein was able to arrive at the same solution for the energy density per frequency interval, $\rho(\nu)$, (equation (6.9)) by using the discrete energy levels of atomic systems. Thus, the quantum of energy $E = h\nu$ is assumed to be the difference between two atomic energy levels,

$$E_2 - E_1 = h\nu. \tag{6.19}$$

Furthermore, Einstein considered that there are only three fundamental processes by which the atomic system can exchange energy with the environment: absorption, spontaneous emission, and stimulated emission (figure 6.7).

Let us consider these three processes separately by denoting the number density of each level by N_1 and N_2. *Absorption* is the process by which an atom in state 1 absorbs an incident photon and is excited to energy level 2. The rate of increase of N_2 due to absorption is proportional to the number density of the atoms in level 1, N_1, and the incident energy density, ρ,

$$\frac{dN_2}{dt} = B_{12} N_1 \, \rho(\nu) \tag{6.20}$$

where B_{12} is the absorption rate, $[B_{12}] = s^{-1}$. As a result of absorption, incident energy is converted into the excitation of atoms from level 1 to 2.

Spontaneous emission is the process whereby an atom from level 2 decays radiatively (with emission of a photon) to the lower state,

$$\frac{dN_2}{dt} = -A_{21} N_2. \tag{6.21}$$

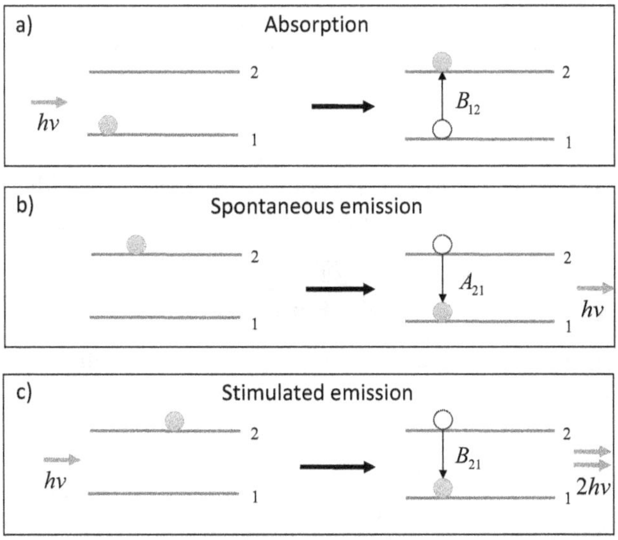

Figure 6.7. Radiative processes in a two-level atomic system: a) absorption, b) spontaneous emission, c) stimulated emission. Note how the stimulated rather than spontaneous emission is the reversed process to absorption.

The coefficient A_{21} is the spontaneous emission rate constant. Note that the inverse of A_{21} can be interpreted as the decay time constant, or *natural lifetime* (see section 5.5), $\tau = 1/A_{21}$.

Stimulated emission is the process by which, in the presence of an incident photon, an excited atom decays to level 1 and releases a photon of the same energy (frequency), direction of propagation, polarization, and phase. Stimulated emission can be regarded as the exact reverse of absorption. This process contrasts with *spontaneous emission*, where the emitted photon has no phase relationship with the stimulating photon, that is, it is emitted with equal probability in all directions of propagation, and with a random direction of the electric field vector (polarization). The rate depends on both the population density in state 2 and the strength of the stimulating light

$$\frac{dN_2}{dt} = -B_{21}N_2\rho(\nu). \tag{6.22}$$

Einstein combined all these processes to express the rate equations,

$$\begin{aligned}\frac{dN_2}{dt} &= -A_{21}N_2 + B_{12}N_1\rho(\nu) - B_{21}N_2\rho(\nu)\\ &= -\frac{dN_1}{dt}.\end{aligned} \tag{6.23}$$

Equation (6.23) also states that the rate of increase in the population of level 2, dN_2/dt, must be accompanied by an identical decrease (hence the negative sign) in the population of level 1, $-dN_1/dt$. Note that here all the nonradiative processes (resulting in loss by heat dissipation) have been ignored. At thermal equilibrium, the excitation and decay mechanisms must balance each other completely, such that

$$\frac{dN_2}{dt} = \frac{dN_1}{dt} = 0. \tag{6.24}$$

Combining equations (6.23) and (6.24), we obtain the ratio of the two population densities,

$$\frac{N_2}{N_1} = \frac{B_{12}\rho(\nu)}{A_{21} + B_{21}\rho(\nu)}. \tag{6.25}$$

Further, Einstein used the classic Boltzmann statistics, which gives the ratio of the two populations as

$$\frac{N_2}{N_1} = \frac{g_2}{g_1} \cdot e^{-\frac{h\nu}{k_B T}}. \tag{6.26}$$

In equation (6.26), quantities g_1 and g_2 are the *degeneracy factors* for the two states, that is, the number of configurations in which a molecule can have the same energy. If we combine equations (6.25) and (6.26) to solve for $\rho(\nu)$, we obtain

$$\rho(\nu) = \frac{A_{21}}{B_{21}} \frac{1}{\frac{B_{12}g_1}{B_{21}g_2} e^{\frac{h\nu}{k_B T}} - 1}. \tag{6.27}$$

By comparing equation (6.27) with Planck's formula (equation (6.9)), Einstein realized that they are identical, provided two conditions are met

$$g_1 B_{12} = g_2 B_{21} \tag{6.28a}$$

$$\frac{A_{21}}{B_{21}} = \frac{8\pi h\nu^3}{c^3}. \tag{6.28b}$$

Equations (6.28*a* and *b*) connect the three Einstein coefficients. Therefore, measurements on a certain radiative process, for example, measuring A_{21}, can inform about the other two coefficients as well as radiation.

6.6 Problems

1. Prove the Stefan–Boltzmann law (equation (6.16)).
2. Prove Wien's displacement law, that is, that the frequency maximum of Planck's curve is proportional to the temperature of the black body, $\nu_{max} \propto T$ (equation (6.15)). Calculate λ_{max}.
3. How many electromagnetic modes exist in a 1 m^3 cavity for light of central wavelength $\lambda_0 = 633$ nm and bandwidth $\Delta\lambda = 1$ nm? What are the central frequency and frequency bandwidth, ν_0, $\Delta\nu$? What are the central angular frequency and angular frequency bandwidth, ω_0, $\Delta\omega_0$?

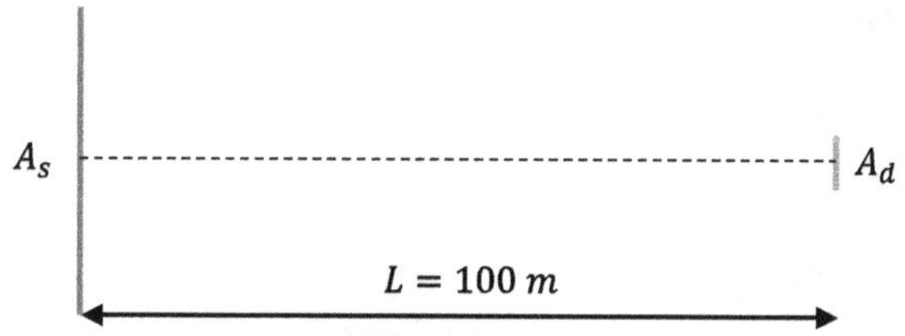

Figure 6.8. Problem 6.7.

4. What is the temperature of a black body whose maximum emission is at $\lambda = 600$ nm, $\lambda = 550$ nm, $\lambda = 400$ nm? Plot the three corresponding spectral emission curves as a function of wavelength, $I(\lambda)$, for all cases.
5. What is the frequency maximum of the Sun's radiation ($T = 5800$ K). If the Sun cooled by 1 F, how much lower would the total (frequency integrated) exitance be?
6. What is the temperature of the black body whose maximum spectral exitance corresponds to the maximum of the photopic curve?
7. A black body source of area $A_s = 10$ cm^2 and temperature $T = 1000$ K emits radiation, which is captured by a photodetector of area $A_D = 1$ cm^2, $L = 100$ m away, as depicted in figure 6.8.

   Consider the source and detector parallel and centered on the same optical axis.

   a) Calculate the power falling on the detector, within the wavelengths interval $\lambda \in [8,12]$ μm.
   b) The detector is moved 10 m farther from the source. What should be the new temperature that will yield the same power, in the same wavelength range, as before?
   c) The detector is moved to a new distance $L_1 = 1$ m ($T = 1000$ K) from the source. What should be the new wavelength interval, centered at $\lambda = 10$ μm, which will result in the same power at the detector as in a)?
8. A black body emits radiation with its maximum spectral exitance, $M_\nu(\nu)$, at frequency $\nu_{\max}$. Express the spectral exitance in terms of wavelength, $M_\lambda(\lambda)$, find the wavelength at maximum, $\lambda_{\max}$, and determine the relationship between $\nu_{\max}$ and $\lambda_{\max}$.
9. Express the spectral exitance of a black body (Planck's formula) in terms of inverse wavelengths, $1/\lambda$.
10. The wavelength-based spectral exitance of a black body is 1 W/(m^2 μm) at its peak. What is the temperature of the source in K?
11. A black body has a temperature of 1000 K and weighs 1 kg.
    a) Assuming constant temperature, how long will it take for the entire mass to be converted into radiation?
    b) The temperature is now $T = 10\ 000$ K. What is the new time of total mass energy conversion?

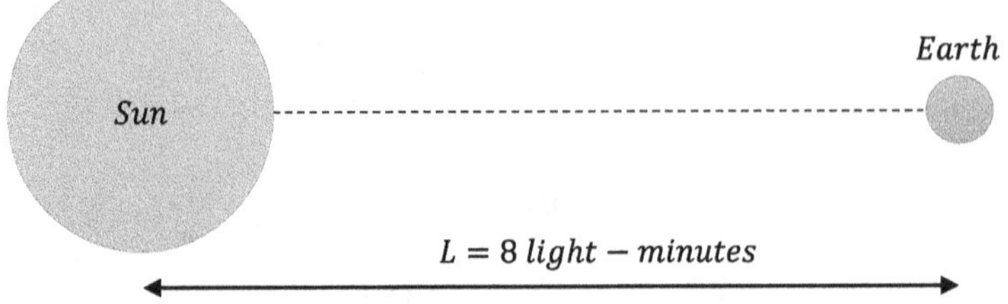

Figure 6.9. Problem 6.13.

12. A black body is cooling at a rate $\alpha = 1$ K/s, starting at $T = 5000$ K. Plot the following versus time:
    a) frequency of maximum emission, $\nu_{\max}$
    b) total exitance, $M = \int_0^\infty M_\nu(\nu)d\nu$
    c) total exitance within the wavelength interval $\lambda \in (1,2)$ μm.
13. Assume the Sun is a black body that floods the Earth's surface with $1\,\text{kW}\,\text{m}^{-2}$ of irradiance (assume isotropic radiation). If the radii of the Earth and Sun are $R_E = 6 \times 10^3$ Km and $R_S = 7 \times 10^5$ Km, respectively, and the distance Sun–Earth is eight light-minutes (see figure 6.9):
    a) What is the total exitance of the Sun?
    b) What is the temperature of the Sun?
    c) What is the total power emitted by the Sun (assuming a spherical shape)?
    d) What is the total power falling on Earth?
    e) If Earth reflects off 20% of the light from the Sun and becomes a Lambertian source, what is the total power received back by the Sun?
14. Plot the following parameters versus temperature $T \in (0,\ \ 10\ 000)$ K for a black body radiator's spectral exitance, $M_\nu(\nu)$.
    a) $\nu_{\max}$, $<\nu>$, and $<\nu> - \nu_{\max}$, where <> denotes the ensemble average over the spectral distribution.
    b) σ_ν, the standard deviation and the spectral variance, $\sigma_\nu = <\nu^2> - <\nu>^2$
    c) skewness, $<\nu^3>$
    d) kurtosis, $<\nu^4>$.
15. Plot the temporal autocorrelation function associated with the black body radiation field at $T = 300$ K.
16. Plot the even and odd components of the black body radiation spectral exitance, respectively,

$$M_\nu^e(\nu) = \frac{M_\nu(\nu) + M_\nu(-\nu)}{2}$$
$$M_\nu^o(\nu) = \frac{M_\nu(\nu) + M_\nu(-\nu)}{2}.$$

Prove that, indeed, M_ν^e and M_ν^o are even and odd, respectively.

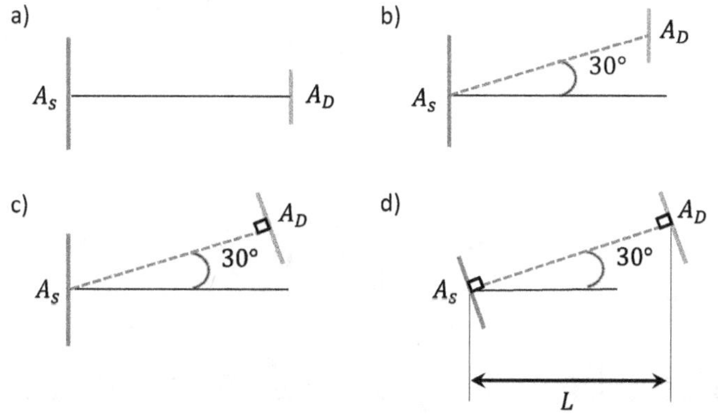

Figure 6.10. Problem 6.19.

17. A flat black body source, of area $A_S = 1\ \text{m}^2$, and temperature $T = 5000$ K, emits radiation that is collected by a detector, of area $A_D = 1\ \text{mm}^2$, at a distance $L = 1$ m.
    a) How much power falls on the detector?
    b) What percentage of the power at the detector is in the visible spectral range, $\lambda \in (380,\ 760)$ nm?
    c) How many photons reach the detector?
18. Derive an expression for the spectral exitance change per unit temperature, assuming a black body source.
19. A Lambertian, black body source and photodetector are placed in four different configurations, as indicated (figure 6.10). If temperature is $T = 3000$ K, $A_s = 1\ \text{cm}^2$, $A_D = 1\ \text{mm}^2$, and $L = 1$ m, compute the power at the detector in all configurations.
20. A normal human body temperature is $T = 37$ °C.
    a) What is the wavelength of maximum spectral exitance, λ_{max}?
    b) By how much does λ_{max} change if the person runs a high fever, $T = 40$ °C?
    c) What is the λ_{max} emitted by an alligator in a pond of temperature $T = 25$ °C?
21. In order to measure the temperature of an unknown object, one experimentalist performs measurements of the black body radiation power within narrow spectral ranges centered at $\lambda_1 = 10$ μm and $\lambda_2 = 20$ μm. It was found that $P_2/P_1 = 0.1$. What is the temperature of the body?

References

[1] Planck M and Masius M 1914 *The Theory of Heat Radiation* (Philadelphia, PA: P. Blakiston's Son & Co.) xiv, p 1

[2] Kingston R H 1978 *Detection of Optical and Infrared Radiation* (Springer Series in Optical Sciences vol 10) (Berlin: Springer), viii p 140

Chapter 7

LASER: light amplification by stimulated emission of radiation

7.1 Population inversion, optical resonator, and narrow band radiation

In 1917, Einstein laid the theoretical foundation for the development of the laser and maser by introducing his probability coefficients (see section 6.5) and re-deriving Planck's formula [1]. A *laser* has the ability to emit light with significant power within a narrow *temporal* and *spatial* frequency range. Since the Fourier transform of the power spectrum is the autocorrelation function (see Volume 1, section 4.3), a narrow *temporal* frequency range implies a long correlation (or coherence) time and a narrow *spatial* frequency range means broad spatial coherence. In other words, laser radiation can be highly *monochromatic* and *collimated.* Note that thermal sources (e.g. incandescent filaments) can be filtered spatially and temporally to approach the collimation and monochromaticity of a laser. However, this procedure is highly *dissipative*, meaning that much of the power is lost in the process. In contrast, laser radiation is emitted in a narrow band spatially and temporally, even in the absence of filtering.

As suggested by the acronym, *stimulated emission* is crucial for laser operation. An atomic system in which the population of the excited level (N_2) is larger than that of the lower level (N_1), $N_2 > N_1$, is said to exhibit *population inversion* (see figures 7.1(a) and (b)). It turns out that this medium can act as an *optical amplifier* (figure 7.1(c)). An atomic system in which population inversion is achieved is sometimes called an *active medium*. Note that at *thermal equilibrium* this situation cannot be achieved, as the Boltzmann statistics require $N_2/N_1 = \exp\left[-(E_2 - E_1)/k_B T\right]$, that is, $N_1 > N_2$. Thus, a laser is an *out-of-equilibrium system*; it can only be constructed by supplying energy from the outside. This process of delivering energy to the atomic system is called *pumping* and can be achieved in a number of ways: electrically, optically, chemically, etc (see figure 7.1).

doi:10.1088/978-0-7503-1644-6ch7

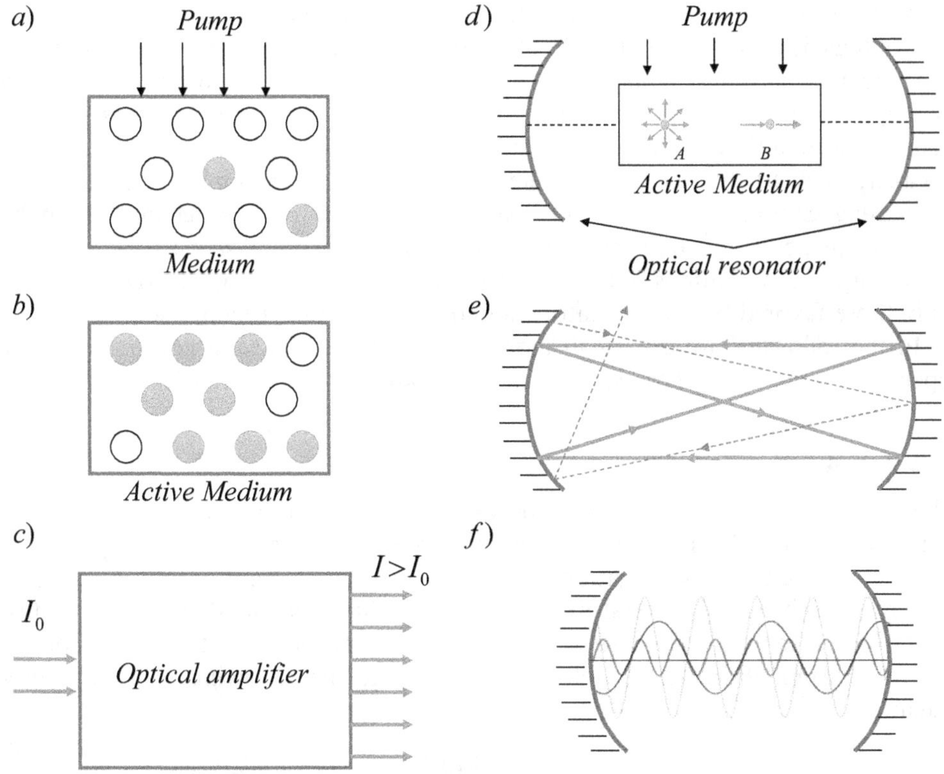

Figure 7.1. (a) Pumping: empty circles represent atoms on ground energy levels and solid circles denote excited atoms. (b) Active medium: population inversion is achieved. (c) Active medium acts as an optical amplifier of the incident radiation. (d) Active medium in a resonator, where A is spontaneous emission, and B is stimulated emission. e) Spatial frequency feedback: the path in green is stable, while the one in red is not (it exits the resonator). (f) Temporal frequency feedback: only the modes with zeros at the boundary survive in the resonator: the mode in the blue line survives, the black and red are suppressed.

Besides the *pumping* required to obtain an *active medium*, the laser requires a mechanism for providing *positive feedback*, that is, for selecting light characterized by narrow spatial and temporal frequency bands to be amplified by the medium (figure 7.1(d)–(f)). This positive feedback mechanism is provided by the optical resonator (cavity), which is essential in converting the pump energy into a small number of *modes:* both longitudinal (temporal frequencies) and transverse (spatial frequencies).

Figure 7.1(d)–(f) illustrates how the *optical resonator*, consisting of two convergent mirrors, provides spatial and temporal feedback. Spontaneous emission (e.g. from atom A in figure 7.1(d)) is isotropically distributed, but the resonator only selects the directions of propagation or wave vector range that can survive a round trip (figure 7.1(e)). Similarly, the cavity favors waves characterized by vanishing electric fields at the boundary. This mechanism selects light of particular wavelengths in the resonator, that is, provides temporal frequency feedback (figure 7.1(f)).

A classical analog of this positive feedback phenomenon is encountered when a microphone is brought close to a speaker: an unpleasant phenomenon occurs, which is called, unsurprisingly, *feedback*. In this case, a very narrow sound frequency band is amplified by the electronic circuit. This particular frequency happens to be a characteristic of the circuit and, following many rounds from the microphone to the speaker, back to the microphone, etc, surpasses all other possible modes in the resonating circuit. An example of spatial, or *transverse mode* feedback, can be observed in a cup of water subjected to some vibrational noise, such as, for instance, on an airplane. Vibrations of broad frequencies are excited at the liquid surface, yet only those favored by the resonator (boundary conditions) become dominant.

In the following, we discuss in more detail several important concepts in laser operation: *gain*, *spectral broadening*, *threshold* and *oscillation*, and *kinetics*.

7.2 Gain

Here we derive an expression for the power gain as the light propagation through an active medium. Let us consider the infinitesimal gain in *spectral irradiance* I_ν, in W/(m^2 Hz) due to the propagation along an element of distance dz (figure 7.2). The incremental change in spectral irradiance, dI_ν, is due to stimulated emission, thus, proportional to I_ν. Further, the amplification is proportional to the volume of the active medium, thus, dz. Therefore, we can write the following proportionality equation,

$$dI_\nu = \gamma(\nu) I_\nu \, dz. \tag{7.1}$$

The proportionality constant γ is called the *gain coefficient*. For small signals, or weak amplification, γ does not depend on I_ν. Physically, this means that the signals are small enough such that the populations of the two levels are not changed significantly due to the incident light (e.g. no saturation). The small signal gain is denoted by γ_0, and equation (2.36) can be readily integrated over the length (L) of the medium to yield

$$I_\nu(L) = I_\nu(0) e^{\gamma_0(\nu) L}. \tag{7.2}$$

Note that if $\gamma_0 < 0$, equation (7.2) becomes the Lambert–Beer law for absorption, denoting an exponential attenuation rather than gain. Now let us find an explicit

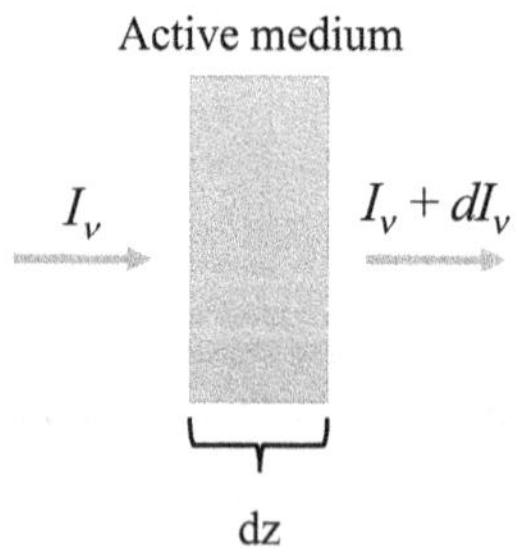

Figure 7.2. Gain through an active medium.

expression for the small signal gain, γ_0, by taking into account the rate equation for the populations of the two levels,

$$\frac{dN_2}{dt} = \int_0^\infty [-A_{21}N_2 g(\nu) - B_{21}N_2\rho(\nu)g(\nu) + B_{12}N_1\rho(\nu)g(\nu)]d\nu. \tag{7.3}$$

In equation (7.3), $g(\nu)$ describes the spectral line shape associated with the transition, $\rho(\nu)$ represents the spectral energy density of the pump, which, as we will see later, is assumed to be much narrower than the line width, $g(\nu)$. The line width, $g(\nu)$, has the meaning of a *probability density*, normalized such that $\int_0^\infty g(\nu)d\nu = 1$. The function $g(\nu)d\nu$ describes the probability of emission of light in the frequency interval $(\nu, \ \nu + d\nu)$.

It has already been discussed (chapter 5) that electronic levels are subdivided into vibrational levels of various energy, each of which can contain a number of rotational levels with different energy. The radiative transitions can take place between various pairs of vibrational and rotational levels. If we denote by $g_{1,2}$ the spectral distribution for each level, then the line shape of the transition takes the form of a correlation between the two functions (figure 7.3),

$$g(\nu) = \int g_1(\nu')g_2(\nu' - \nu)d\nu'. \tag{7.4}$$

Therefore, using the 'moment of the convolution/correlation theorem' (Volume 1, section 4.3) the bandwidth, $\Delta\nu_{12}$, defined as the standard deviation of $g(\nu)$, satisfies (see equation (4.42), Volume 1)

$$\begin{aligned}
\Delta\nu_{12}^2 &= \langle\nu_{12}^2\rangle - \langle\nu_{12}\rangle^2 \\
&= \langle\nu_1^2\rangle + \langle\nu_2^2\rangle + 2\langle\nu_1\rangle\langle\nu_2\rangle - \langle\nu_1 + \nu_2\rangle^2 \\
&= \langle\nu_1^2\rangle - \langle\nu_1\rangle^2 + \langle\nu_2^2\rangle - \langle\nu_2\rangle^2 \\
&= \Delta\nu_1^2 + \Delta\nu_2^2.
\end{aligned} \tag{7.5}$$

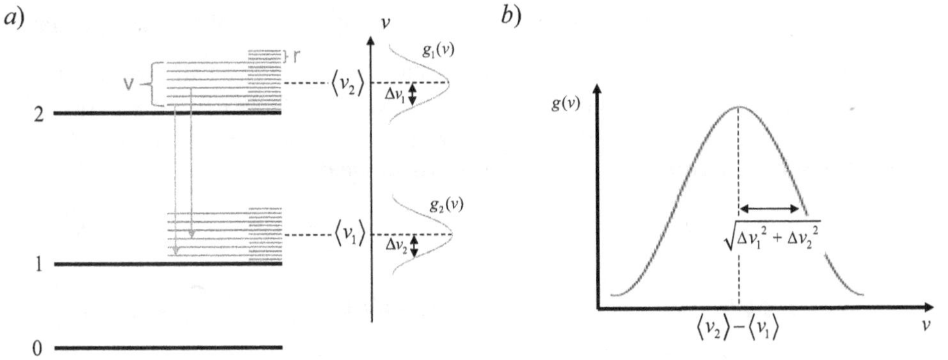

Figure 7.3. (a) Electronic states, 0, 1, 2, each containing vibrational level (v, blue lines) and respective rotational levels (r, red lines). The energy distributions of the upper levels 1 and 2 are g_1, g_2. (b) The transition lineshape, g, is the cross-correlation function of g_1 and g_2, with the average and standard deviation as indicated.

In equation (7.5), $\langle \nu \rangle_{1,2}$ denotes, as usual, the mean frequency of the distribution $g_{1,2}$, $\Delta\nu_{1,2}^2$ are their variances, and $\Delta\nu_{12}^2$ is the variance of g. It is left as an exercise to show that the average of the correlation function, g, is $\langle \nu \rangle = \langle \nu_2 \rangle - \langle \nu_1 \rangle$. The emission spectrum is broader than either of the individual levels, as the variances add up. As we will see shortly, certain spectral distributions, for example, the Lorentzian, have second moments that diverge. This is related to the fact that the Fourier transform of the Lorentz function, the double exponential, $\exp(-|\tau|/\tau_0)$, is not derivable at the origin. The standard deviation is not a good measure of bandwidth in this case.

Note that the line shape $g(\nu)$ describes not only the probability of emitting a photon within the frequency interval $(\nu, \nu + d\nu)$, via spontaneous or stimulated emission, but also the probability of absorption within the same interval. This becomes clear if we recall that absorption and stimulated emission are reverse processes of one another.

If we multiply equation (7.3) by $h\nu$, we obtain the energy radiated per unit time and volume, or $d^2Q/(dt\, dV) = d^2Q/(dt\, A\, dz)$, where A is the area normal to direction z and $A\, dz = dV$ defines the infinitesimal volume of interest. Since *irradiance* is power per unit area, $d^2Q/(dt\, dA)$, we can write its change with respect to the distance as

$$\begin{aligned}\frac{dI_\nu}{dz} &= \frac{d^2Q}{A\, dt\, dz}\\ &= \frac{d^2Q}{dt\, dV}\\ &= h\nu\frac{dN_2}{dt}\\ &= -A_{21}N_2 - \int_0^\infty [B_{21}N_2\rho(\nu)g(\nu) - B_{12}N_1\rho(\nu)g(\nu)]d\nu,\end{aligned} \tag{7.6}$$

where the spontaneous emission term on the right-hand side (RHS), $-A_{21}N_2 = -A_{21}N_2\int_0^\infty g(\nu)d\nu$, because $\int_0^\infty g(\nu)d\nu = 1$. To simplify the calculation of the integrals on the RHS, we will assume that the pump spectral density, $\rho(\nu)$, is much narrower than the absorption or emission line width, $g(\nu)$ (see section 7.4 in [2]). Thus, if we approximate $\rho(\nu') \simeq \rho(\nu)\delta(\nu - \nu')$, the two integrals become convolutions with a delta function and thus, disappear

$$\begin{aligned}&\int_0^\infty [B_{21}N_2\rho(\nu')g(\nu') - B_{12}N_1\rho(\nu')g(\nu')]d\nu'\\ &= \int_0^\infty [B_{21}N_2\rho(\nu)\delta(\nu - \nu')g(\nu') - B_{12}N_1\rho(\nu)\delta(\nu - \nu')g(\nu')]d\nu'\\ &= B_{21}N_2\rho(\nu)g(\nu) - B_{12}N_1\rho(\nu)g(\nu).\end{aligned} \tag{7.7}$$

In order to simplify equation (7.7), we invoke the relationships between Einstein's coefficients, that is, equations (6.28*a*–*b*), and write the result in terms of A_{21} alone,

$$\frac{1}{h\nu}\frac{dI_\nu}{dz} = -A_{21}N_2 - A_{21}\frac{c^3}{8\pi h\nu^3}\rho(\nu)g(\nu)\left[N_2 - \frac{g_2}{g_1}N_1\right]. \tag{7.8}$$

In equation (7.8), $g_{1,2}$ are, as before, the degeneracies of the two levels. Using the relationship between the energy density $\rho(\nu)$ and spectral irradiance, I_ν, for propagation along the z-direction, that is $I_\nu = \rho(\nu)c$, we can write equation (7.8) as

$$\frac{1}{h\nu}\frac{dI_\nu}{dz} = -A_{21} \cdot N_2 - \frac{\sigma(\nu)I_\nu}{h\nu}\Delta N. \tag{7.9}$$

In equation (7.9), σ and ΔN are referred to as the *stimulated emission cross section* and *population inversion*, respectively, defined as

$$\begin{aligned}\sigma(\nu) &= A_{21}\frac{c^2}{8\pi\nu^2}g(\nu) = A_{21}\frac{\lambda^2}{8\pi}g(\nu)\\ \Delta N &= N_2 - \frac{g_2}{g_1}N_1.\end{aligned} \tag{7.10}$$

The quantity σ has units of area, hence the name cross section. σ quantifies how much stimulated emission power (in W) is emitted by each atom exposed to a certain irradiance (W/m^2). Note that the spontaneous emission term, $-A_{21}N_2$, represents *noise* in the laser system.

The laser line emerges from the noise as the result of the positive feedback provided by the resonator. Therefore, during normal laser operation, the spontaneous emission term becomes subdominant and can be neglected. With this approximation, we re-write equation (7.10) as

$$\begin{aligned}\frac{dI_\nu}{dz} &= \gamma(\nu)I_\nu\\ &= \sigma(\nu)I_\nu\Delta N\end{aligned} \tag{7.11a}$$

where

$$\begin{aligned}\gamma(\nu) &= A_{21}\frac{\lambda^2}{8\pi}g(\nu)\left[N_2 - \frac{g_2}{g_1}N_1\right]\\ &= \sigma(\nu)\Delta N.\end{aligned} \tag{7.11b}$$

Equations (7.11*a*–*b*) are the main result of this section as the gain, γ, is fundamental to laser operation. As mentioned already, N_1 and N_2 can depend on I_ν, which means that γ depends on I_ν. As an example, consider that as I_ν continues to increase via amplification, eventually the population of the excited state, N_2, decreases ('depletes') due to radiative decay and approaches N_1, that is, the gain reduces or *saturates*. Gain saturation effects must include careful consideration of the pump, as further discussed in the next section.

For now, let us consider the small signal gain, γ_0, which is independent of I_ν. Importantly, the gain coefficient is positive only if there is a population inversion, that is, if $\gamma > 0$, then $\Delta N > 0$. If $\Delta N < 0$, then $\gamma < 0$ and $|\gamma|$ denotes the absorption coefficient. For positive gain, equation (7.2) is the integral over the z form of equation (7.11a), and can be further expressed as

$$I_\nu(L) = I_\nu(0)G_0(\nu, L) \tag{7.12}$$

where G_0 is called the (small signal) *power gain,* defined as (see equation (7.2))

$$G_0(\nu, L) = e^{\gamma_0(\nu)L}. \tag{7.13}$$

Unlike the gain coefficient, γ_0, the power gain depends on the length of the medium. Significantly, G_0 acts as a frequency filter on the initial irradiance I_ν. Because of the exponential dependence on $\gamma_0(\nu)$ and, thus, $g(\nu)$, $G_0(\nu)$ is a narrower function of frequency (see figure 7.4). Therefore, G_0 is a *band-pass filter*, which narrows down progressively as L increases, that is, as light bounces back and forth for longer in the resonator.

We can understand this narrowing of the line with longer propagation by invoking the uncertainty relation applied to temporal signals (Volume 1, chapter 8): longer propagation in the medium is equivalent to a longer signal spread in time. Due to a time–frequency uncertainty, this results in a narrower spatial distribution, because $\Delta\omega \geqslant \dfrac{1}{2\Delta t}$.

We can conclude that the emitted laser light depends not only on the line shape of the transition, $g(\nu)$, a property of the gain medium, but also on the resonator properties. In the following section, we discuss the factors that affect the spectral line. The shape and width of the emission line have important consequences on the coherence properties of the light and certain applications to imaging.

7.3 Spectral line broadening

According to the *uncertainty principle* (Volume 1, chapter 8), the line width of any transition has a finite bandwidth, simply because the light transit in the cavity is finite in time. There are many factors that can contribute to broadening this spectral line. The broadening mechanisms are classified in two groups: *homogeneous broadening*, when the mechanism is the same for all atoms, and *inhomogeneous broadening*, when different subgroups of atoms broaden the line differently.

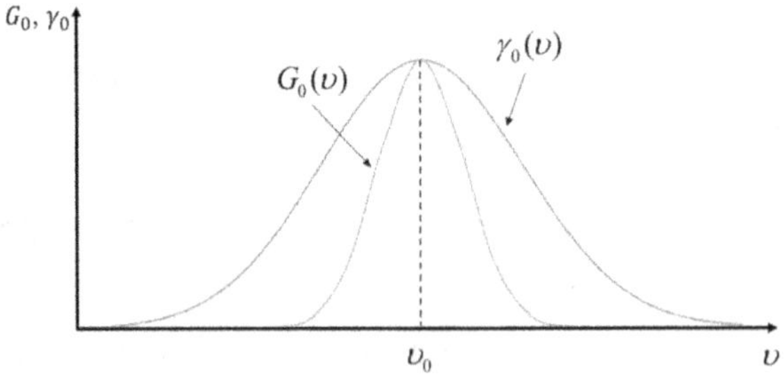

Figure 7.4. The (small signal) power gain, G_0, is narrower in frequency than the gain coefficient, γ_0.

7.3.1 Homogeneous broadening

- *Natural (lifetime) broadening*

  This broadening is due to the lifetime of the excited state in the absence of nonradiative decay, that is, due to the *natural lifetime* (see previous discussion in section 5.5). This is the lifetime of the atoms that are not disturbed by other processes (e.g. collisions with other atoms), which have the effect of shortening this lifetime. As mentioned earlier, the natural lifetime is simply the inverse of the spontaneous emission rate (see equation (5.7)),

$$\frac{1}{\tau_{21}} = A_{21}. \tag{7.14}$$

In other words, the natural lifetime is the largest possible; conversely, the natural line shape is the narrowest. The natural line shape is a *Lorentzian* centered (shifted) at $\nu_0 = \langle \nu_2 \rangle - \langle \nu_1 \rangle$, of half-width at half maximum $\Delta\nu/2$ (see figure 7.5),

$$g(\nu) = \frac{1}{\pi}\frac{\Delta\nu/2}{1 + \frac{(\nu - \nu_0)^2}{(\Delta\nu/2)^2}}. \tag{7.15}$$

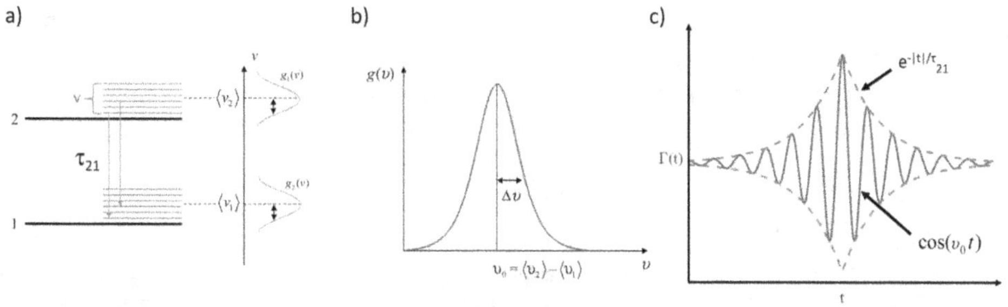

Figure 7.5. (a) Transition between two levels, with the lifetime as indicated. (b) Spectral line of the emission. (c) Temporal autocorrelation function associated with spontaneous emission, i.e., the Fourier transform of the power spectrum in (b).

It can be seen that this line width is normalized to the unit area, $\int_0^\infty g(\nu)d\nu = 1$.

The temporal behavior of *spontaneous emission* can be described by the inverse Fourier transform of $g(\nu)$, which yields a temporal autocorrelation function of the form (using the Fourier transform of a Lorentzian and shift theorem, see chapter 4 in Volume 1)

$$\Gamma(t) = e^{-i2\pi\nu_0 t}e^{-2\pi\Delta\nu t}. \tag{7.16}$$

Equation (7.16) describes an exponentially damped oscillation (figure 7.5(c)). It is left as an exercise to show that for a Lorentzian distribution, the standard deviation diverges and, thus, cannot be used as a measure of bandwidth.

- *Collision broadening*

  The process of collisions between atoms is essential in gas lasers. The higher the pressure of the laser tube, the more important this effect becomes, resulting in line broadening. Sometimes this phenomenon is referred to as *pressure broadening*.

  Let us consider that, on average, each atom experiences collisions at frequency ν_{col}. The net effect of these collisions is to interrupt the emission process, essentially shortening the lifetime of both levels, 1 and 2, by the same amount,

$$\begin{aligned} A_{21}^{col} &= A_{21} + \nu_{col} \\ A_{10}^{col} &= A_{10} + \nu_{col}. \end{aligned} \tag{7.17}$$

  Therefore, the collision-broadened line width is

$$\Delta\nu_{col} = \frac{1}{2\pi}[A_{21} + A_{10} + 2\nu_{col}]. \tag{7.18}$$

  In many situations of practical interest, the frequency of collision is much higher than both the natural line widths, $\nu_{col} \gg A_{21}, A_{10}$, such that we can approximate the collision-induced bandwidth by twice the collision frequency,

$$\Delta\nu_{col} \simeq \nu_{col}/\pi. \tag{7.19}$$

  Typically, collision broadening is expressed as MHz of broadening per unit of pressure. Since the frequency of collisions is assumed to be the same for all atoms, this type of broadening is indeed *homogeneous*. In solids, pressure broadening is absent.

7.3.2 Inhomogeneous broadening

Groups of atoms can differ from one another in terms of their velocities, masses, nuclear spins, etc. These differences produce *inhomogeneous* line broadening, such as Doppler broadening (velocity related), isotope broadening (mass related), Zeeman splitting (nuclear spins in magnetic fields), or Stokes splitting (nuclear spins in electric fields).

- *Doppler broadening*

  At thermal equilibrium, the atoms in a gas have a distribution of velocities along one axis (say, z), v_z, that obeys the Maxwell–Boltzmann statistics, that is, a Gaussian distribution,

$$p(\nu_z) = \frac{1}{\sqrt{2\pi}\,\Delta\nu} \cdot e^{-\frac{M\nu_z^2}{2k_BT}} = \frac{1}{\sqrt{2\pi}\,\Delta\nu} \cdot e^{-\frac{\nu_z^2}{2(\Delta\nu)^2}} \tag{7.20}$$

where M is the atomic mass, T is the absolute temperature, and k_B is the Boltzmann constant. The standard deviation of the Gaussian distribution is

$$\Delta\nu = \sqrt{\frac{k_BT}{M}}. \tag{7.21}$$

Note that the probability density is normalized such that $\int_{-\infty}^{\infty} p(\nu_z)d\nu_z = 1$. During the radiative decay, a group of atoms moving with velocity ν_z emits light at a Doppler frequency, ν_D, that is shifted with respect to the rest ($\nu_z = 0$) frequency, ν_0, by

$$\nu_D - \nu_0 = \frac{\nu_z}{c}\nu_0. \tag{7.22}$$

As a result, the *natural line* described earlier (equation (7.15)) must be corrected for the Doppler effect as

$$g\left(\nu - \nu_0\frac{\nu_z}{c}\right) = \frac{1}{2\pi}\frac{\Delta\nu_h}{1 + \left(\frac{\nu - \nu_0 - \nu_0\frac{\nu_z}{c}}{\frac{\Delta\nu_h}{2}}\right)^2}. \tag{7.23}$$

Averaging these Doppler shifts for all atoms in the system requires integrating over all velocities (note that the velocity component ν_z has both positive and negative values). The result is known as the *inhomogeneous Doppler line broadening*, which is the *convolution* between the Lorentzian associated with the homogeneous line, and the Gaussian velocity distribution (see figure (7.6))

$$g_D(\nu) = \int_{-\infty}^{\infty} p(\nu_z)g\left(\nu - \nu_0\frac{\nu_z}{c}\right)d\nu_z = p \circledcirc g. \tag{7.24}$$

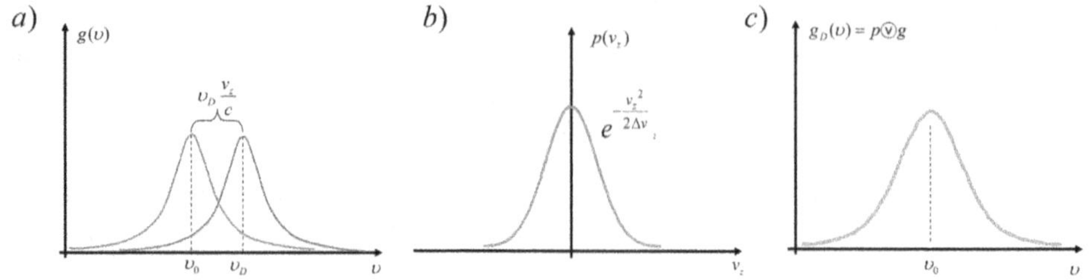

Figure 7.6. (a) Doppler frequency shift of the 'natural' transition line, for a single velocity ν_z. (b) Maxwell–Boltzmann distribution of velocities. (c) Voigt distribution: the Doppler-broadened spectral line, averaged over all velocities.

This function is known in the laser community as the *Voigt distribution*. This convolution can be Fourier transformed to the time domain, where it becomes a product between an exponential (Fourier transform of the Lorentzian) and a Gaussian (Fourier transform of the Gaussian in equation (7.20)),

$$\tilde{g}_D(t) \propto e^{-a|t|-bt^2} e^{i\upsilon_0 t}. \tag{7.25}$$

In equation (7.25), we ignored the normalization factor and introduced constants a and b just to show the functional dependence with respect to time. The Voigt profile, $g_D(\nu)$ reduces to p if g approaches a δ-function, in which case the Doppler broadening is dominant over the natural broadening. Doppler broadening is reduced in low-temperature gases and, of course, in solid state lasers it is absent (p approaches a delta function).

It is left as an exercise to study the problem when the atomic system has a non-zero *average velocity*, such as in the case of gas flow (see problem 4). In this case the velocity distribution p is shifted to $<\nu_z>$ and the 'moment of a convolution' property (Volume 1, section 4.3) provides rapid insight into the solution.

- *Isotopic broadening*

  Sometimes the active medium contains *isotopes* of the same chemical element, that is, variants of atoms that contain a different number of neutrons in the nucleus and, thus, different masses. One common example is the helium–neon (He–Ne) laser, as neon occurs naturally in two isotopes, ^{20}Ne and ^{22}Ne. The spectral line is slightly shifted depending on which isotope emits light.
- *Stokes and Zeeman splitting*

  Stokes and Zeeman effects describe the splitting of the line due to, respectively, the electric and magnetic fields present. These phenomena are important in solid state lasers, where atoms in the lattice are exposed to inhomogeneous (local) fields.

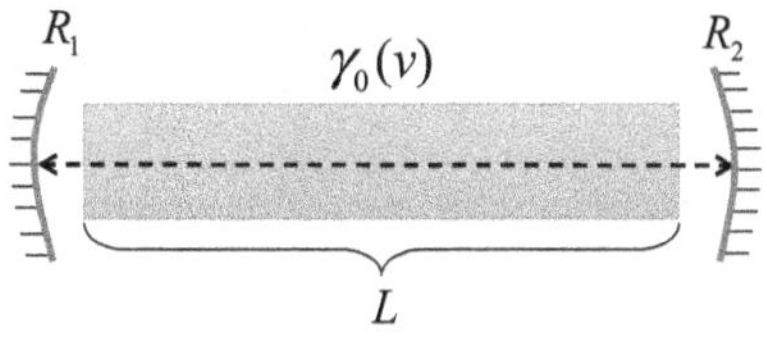

Figure 7.7. Losses in the resonator due to mirror reflection.

7.4 Threshold for laser oscillation

In section 7.2 we discussed the weak signal laser gain, $\gamma_0(\nu) = \sigma(\nu)\Delta N$ (recall equations (7.11*a–b*). As long as there is population inversion, $\Delta N > 0$, the active medium operates as an amplifier. Yet, this condition is not sufficient for *laser oscillation*, because the resonator introduces losses which may overwhelm the gain. The simplest kind of losses is due to reflections at the resonator mirrors (figure 7.7).

The threshold condition requires that, upon a round trip propagation in the resonator, the irradiance experiences a net gain,

$$e^{2\gamma_0(\nu)L} R_1 R_2 \geqslant 1. \tag{7.26}$$

From equation (7.26), we can obtain the minimum, or *threshold* gain coefficient that allows laser oscillation,

$$\gamma_{th} = -\frac{1}{2L} \ln(R_1 R_2). \tag{7.27}$$

Note that the threshold condition acts as a filter in the frequency domain, because frequencies at the center of the line are amplified better and, thus, are more likely to oscillate than frequencies on the wings of the line. For calculating the threshold gain, the stimulated emission can be ignored. However, as the power increases, stimulated emission takes over and, eventually, spontaneous emission can be ignored. For very high gains, saturation effects may become important. Next, we study how lasing behaves as a function of time.

7.5 Laser kinetics

The time evolution of laser radiation is dictated by the kinetics of the population in each level. It has been shown already that a two-level system is incompatible with laser radiation (see section 7.1 for this discussion). Here we discuss a generic 'three-level system' and problems related to four-level systems are left as an exercise. Let us consider the diagram shown in figure 7.8, where the radiative transition of interest takes place between levels 2 and 1, while 0 denotes the ground state.

The rate equations for the upper two levels of interest can be written as

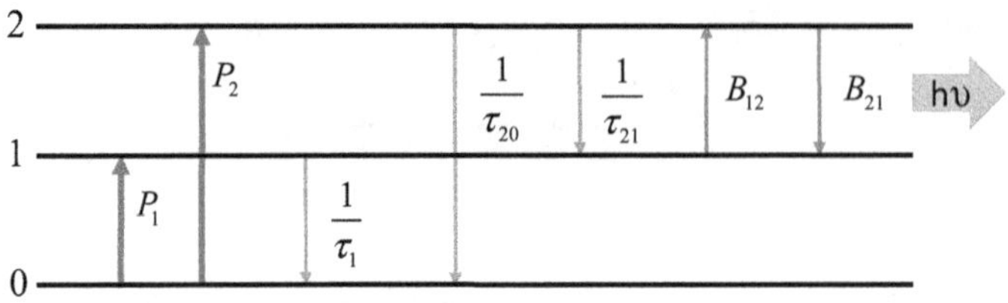

Figure 7.8. Three-level system: laser emission takes place due to radiative transition between levels 2 and 1.

$$\frac{dN_2(t)}{dt} = P_2(t) - \frac{N_2(t)}{\tau_2} - \frac{\sigma(\nu)I(\nu)}{h\nu}[N_2(t) - N_1(t)] \tag{7.28a}$$

$$\frac{dN_1(t)}{dt} = P_1(t) - \frac{N_1(t)}{\tau_1} + \frac{N_2(t)}{\tau_{21}} + \frac{\sigma(\nu)I(\nu)}{h\nu}[N_2(t) - N_1(t)]. \tag{7.28b}$$

Equations (7.28*a*–*b*) are coupled linear differential equations, with the following parameters:

- P_1 and P_2 are the pump rates of levels 1 and 2, respectively (we assume the ground state, N_0, never depletes);
- $\frac{1}{\tau_2} = \frac{1}{\tau_{20}} + \frac{1}{\tau_{21}}$ is the inverse lifetime (decay rate) of level 2, $\frac{1}{\tau_{20}}$ the decay rate to the ground state, and $\frac{1}{\tau_{21}}$ the rate of spontaneous emission;
- σ is the spontaneous emission cross section;
- $I(\nu)$ is the spectral irradiance;
- $N_2 - N_1$ is the population inversion (assuming equal degeneracies, $g_1 = g_2$);
- τ_1 is the lifetime of level 1.

Solving for $N_1(t)$ and $N_2(t)$ in equations (7.28*a*–*b*) can be very efficiently performed by using the Laplace transform (see Volume 1, chapter 11, for a review of the Laplace transform). The Laplace transform is a powerful tool for studying *transient phenomena* such as population dynamics in an atomic system. These phenomena occur in *initial value problems*, whenever the system under investigation has a 'switch-on or -off'. In equations (7.28*a*–*b*), the pumps $P_{1,2}$ turn on at a particular moment in time. In the simplest case, we can consider $P_{1,2}$ as being *turned on* at $t = 0$ and staying constant for $t > 0$, that is, described by a step function,

$$P_{1,2}(t) = \begin{vmatrix} P_{1,2}, & t > 0 \\ \frac{1}{2}P_{1,2}, & t = 0 \\ 0, & \text{rest} \end{vmatrix} \tag{7.29}$$
$$= P_{1,2}\Gamma(t),$$

where $\Gamma(t)$ is the Heaviside (step) function. Now we recall the Laplace transform of the step function as well as the differential operator,

$$\begin{aligned} \Gamma(t) &\leftrightarrow \frac{1}{s} \\ \frac{df(t)}{dt} &\leftrightarrow s\tilde{f}(s), \end{aligned} \tag{7.30}$$

where the double arrow indicates the Laplace transform operation. These properties allow us to transform the differential equations (7.28*a*–*b*) into an algebraic equation,

$$s\tilde{N}_2(s) = \frac{P_2}{s} - \frac{\tilde{N}_2(s)}{\tau_2} - \frac{\sigma I}{h\nu}[\tilde{N}_2(s) - \tilde{N}_1(s)] \tag{7.31a}$$

$$s\tilde{N}_1(s) = \frac{P_1}{s} - \frac{\tilde{N}_1(s)}{\tau_1} + \frac{\tilde{N}_2(s)}{\tau_{21}} + \frac{\sigma I}{h\nu}[\tilde{N}_2(s) - \tilde{N}_1(s)]. \tag{7.31b}$$

This system of equations can now be easily solved in the Laplace (s) space; the only challenge remaining is to transform the final result back into the time domain. Various particular cases can be solved, depending on which terms in equation (7.31) can be safely neglected, as follows:

- $P_1 = 0$, no pumping of level 1
- $\frac{1}{\tau_2} = 0$, no spontaneous decay of level 2 (large signal gain)
- $\sigma = 0$, no stimulated emission (weak signal)
- $\frac{1}{\tau_{21}} = 0$, no spontaneous emission
- $\frac{dN_1}{dt} = \frac{dN_2}{dt} = 0$, steady state.

Many combinations of these particular situations are left as exercises (see section 7.7 problems).

Example 1: $P_1 = 0$, $\sigma = 0$.

Here we illustrate the procedure of solving the rate equations, using for simplicity $P_1 = 0$, $\sigma = 0$. In this case the solutions of equation (7.31) become

$$\begin{aligned} \left(s + \frac{1}{\tau_2}\right)\tilde{N}_2(s) &= \frac{P_2}{s} \\ \left(s + \frac{1}{\tau_1}\right)\tilde{N}_1(s) &= \frac{\tilde{N}_2(s)}{\tau_{21}}. \end{aligned} \tag{7.32}$$

$\tilde{N}_2(s)$ can be rearranged using partial fraction decomposition (or expansion), as

$$\tilde{N}_2(s) = \frac{P_2}{s\left(s + \dfrac{1}{\tau_{21}}\right)} = P_2\tau_2\left[\frac{1}{s} - \frac{1}{s + 1/\tau_2}\right]. \tag{7.33}$$

In order to arrive at $N(t)$ we use the known Laplace pair, $\frac{1}{s} \leftrightarrow 1$ and the *shift property* of the Laplace transform (see Volume 1, section 11.1),

$$\tilde{f}(s + a) \rightarrow e^{-at}f(t). \tag{7.34}$$

Using this property, we take the Laplace transform inverse of equation (7.33) and readily obtain

$$N_2(t) = P_2\tau_2\left(1 - e^{-\frac{t}{\tau_2}}\right). \tag{7.35}$$

The population dynamics of level 2 are depicted in figure 7.9 for two ratios of the lifetimes.

Solving for N_1 is slightly more tedious because it depends on two lifetimes, τ_{21} and τ_1. Still, the procedure is the same, expressing $N(s)$ as a sum of *simple fractions*, of the form $\frac{a}{s + b}$, which can be easily Laplace-transformed. From equation (7.32b), we have $N_1(s)$ as

$$N_1(s) = \frac{P_2}{\tau_{21}} \frac{1}{s\left(s + \dfrac{1}{\tau_2}\right)\left(s + \dfrac{1}{\tau_1}\right)}. \tag{7.36}$$

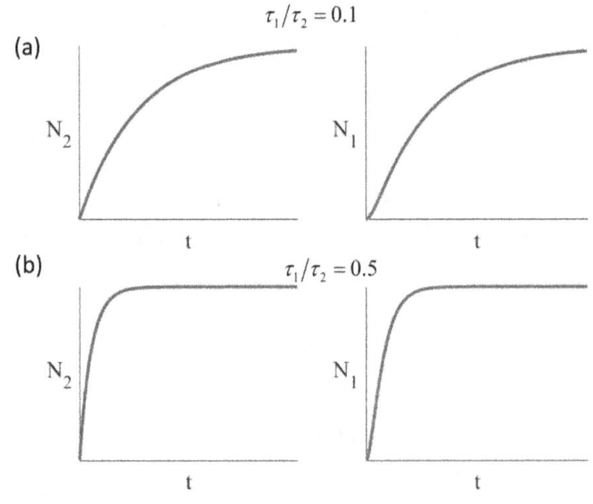

Figure 7.9. Population dynamics for the two levels expressed in equations (7.35) (N_2) and (7.39) (N_1) for two ratios of the lifetimes, as indicated.

We can further express $N_1(s)$ as

$$N_1(s) = \frac{P_2}{\tau_{21}}\left[\frac{A}{s} + \frac{B}{s + \frac{1}{\tau_2}} + \frac{C}{s + \frac{1}{\tau_1}}\right], \tag{7.37}$$

where the constants A, B and C can be identified by comparing equations (7.36) and (7.37). Thus, we find

$$A\left(s + \frac{1}{\tau_2}\right)\left(s + \frac{1}{\tau_1}\right) + B \cdot s\left(s + \frac{1}{\tau_1}\right) + C \cdot s\left(s + \frac{1}{\tau_2}\right) = 1.$$

As a result, the three constants are found easily by multiplying both sides of the equation to isolate one constant, say, A, and then evaluating the expression at a value of s that vanishes the terms containing the other two constants, say B and C. Thus,

$$A = \left.\frac{1}{\left(s + \frac{1}{\tau_2}\right)\left(s + \frac{1}{\tau_1}\right)}\right|_{s=0} = \tau_1\tau_2 \tag{7.38a}$$

$$B = \left.\frac{1}{s\left(s + \frac{1}{\tau_1}\right)}\right|_{s+\frac{1}{\tau_2}=0} = \tau_1\tau_2\frac{1}{\frac{\tau_1}{\tau_2} - 1} \tag{7.38b}$$

$$C = \left.\frac{1}{s\left(s + \frac{1}{\tau_2}\right)}\right|_{s+\frac{1}{\tau_1}=0} = \tau_1\tau_2\frac{\tau_1/\tau_2}{1 - \frac{\tau_1}{\tau_2}}. \tag{7.38c}$$

The time domain solution is obtained at once by taking the Laplace transform inverse of equation (7.37) and using the A, B, C constants from equations (7.38a–c),

$$N_1(t) = P_2\frac{\tau_1\tau_2}{\tau_2}\left[1 + \frac{\tau_1/\tau_2}{1 - \tau_1/\tau_2}e^{-\frac{t}{\tau_1}} - \frac{1}{1 - \tau_1/\tau_2}e^{-\frac{t}{\tau_2}}\right]. \tag{7.39}$$

The trend of $N_1(t)$ for two values of the ratio τ_1/τ_2 is shown in figure 7.9. Clearly, if $\tau_2 > \tau_1$, the density N_2 is greater than N_1 and gain is obtained across all values of t. However, if $\tau_1 > \tau_2$, the excited level is depleted and gain quickly diminishes. Note that the two steady states are obtained as the limit of $N_1(t)$, $N_2(t)$, for $t \to \infty$,

$$\begin{aligned} N_1(t \to \infty) &= P_2\frac{\tau_1\tau_2}{\tau_{21}} \\ N_2(t \to \infty) &= P_2\tau_2. \end{aligned} \tag{7.40}$$

7.5.1 Partial fraction decomposition

Note that the (partial fraction) expansion in equations (7.33) and (7.36) is very useful, as the individual terms resulting from the expansion have simple Laplace transforms (exponentials). Whenever a function is in the form of a ratio of two polynomial functions, $f(s) = g(s)/h(s)$, g and h with no common factors and g of lower degree than h, it is convenient to expand f as (for a review on partial fraction decomposition, see, e.g. p 942 in [3])

$$f(s) = \sum_n \frac{b_n}{s - a_n},$$

where a_n are the roots of h and b_n are constants that result from the identity $g(s)/h(s) = \sum_n \frac{b_n}{s - a_n}$, which can be found as described in equation (7.37) for coefficients A, B, and C. Note how quickly we can now perform the Laplace transform of f, using the shift theorem,

$$f(t) = \sum_n b_n e^{a_n t}.$$

If a certain root, say a_1, has a multiplicity m, that is, there is a fraction of the form $\frac{1}{(s - a_1)^m}$, the partial fraction expansion becomes

$$f(s) = \sum_{n \neq 1} \frac{b_n}{s - a_n} + \sum_{k=1}^{m} \frac{b_{1k}}{(s - a_1)^k}.$$

In this case, we invoke the shift theorem in combination with the known Laplace transform pair (see table 11.1 in Volume 1, chapter 11)

$$\frac{1}{s^n} \leftrightarrow \frac{t^{n-1}}{(n-1)!}.$$

Thus, we arrive at the following result that can be used whenever the polynomial in the denominator has roots of multiplicity greater than one, namely,

$$\frac{b_{1k}}{(s-a_1)^k} \leftrightarrow \frac{b_{1k}}{(k-1)!}t^{k-1}e^{a_1 t}.$$

7.6 Gain saturation

Let us now return to the general rate equations (7.28*a–b*) and their Laplace transforms (equations (7.31*a–b*), and now investigate the effect of stimulated emission ($\sigma \neq 0$). For simplicity, let us assume $\tau_1 = 0$, which means that level 1 decays infinitely fast, that is, $N_1 = 0$. Under these circumstances, equation (7.31*a*) for the population dynamics of level 2 in the Laplace domain, becomes

$$\tilde{N}_2(s)\left[s + \frac{1}{\tau_2} + \frac{\sigma I}{h\nu}\right] = \frac{P_2}{s}. \tag{7.41}$$

It is quite informative to introduce a new lifetime, τ_2', adjusted for the effects of stimulated emission,

$$\begin{aligned}\frac{1}{\tau_2'} &= \frac{1}{\tau_2}\left[1 + \frac{\sigma I \tau_2}{h\nu}\right] \\ &= \frac{1}{\tau_2}\left(1 + \frac{I}{I_s}\right),\end{aligned} \tag{7.42}$$

which brings equation (7.41) to the same form as equation (7.32). In equation (7.42), $I_s = h\nu/\sigma\tau_2$ is called *saturation irradiance*. Transforming back to the time domain, we obtain a solution similar to that of equation (7.35),

$$N_2(t) = P_2\tau_2'\left(1 - e^{-\frac{t}{\tau_2'}}\right). \tag{7.43}$$

If the irradiance I becomes comparable with I_s, the lifetime τ_2' is significantly shortened with respect to the original τ_2 (for $I = I_s$, $\tau_2' = \tau_2/2$). A shorter lifetime means faster depletion of level 2. Thus, not surprisingly, whenever I approaches I_s,

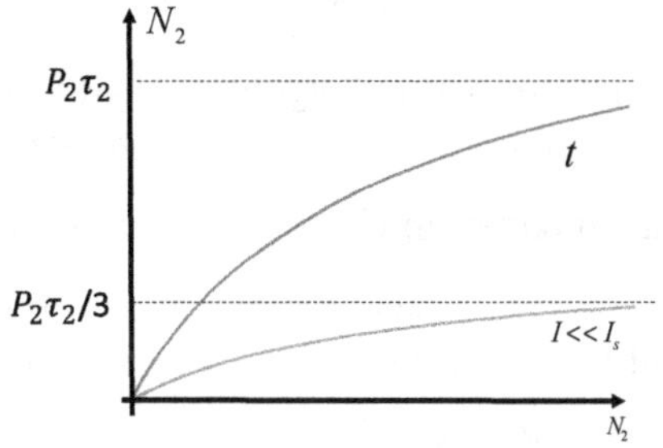

Figure 7.10. Effects of the saturation intensity on the N_2 kinetics.

the gain *saturates*. Figure 7.10 shows the general behavior of $N_2(t)$ for $I \ll I_s$ and $I = 2I_s$.

7.6.1 Saturation at steady state

Let us solve the general equations (equation (7.28)) at *steady state*, that is, when $\frac{dN_1}{dt} = \frac{dN_2}{dt} = 0$,

$$\begin{aligned} P_2 - \frac{N_2}{\tau_2} - \frac{\sigma I}{h\nu}(N_2 - N_1) &= 0 \\ P_1 + \frac{N_2}{\tau_{21}} + \frac{\sigma I}{h\nu}(N_2 - N_1) - \frac{N_1}{\tau_1} &= 0. \end{aligned} \tag{7.44}$$

After some simple (but tedious) calculations, these equations give a population difference, $N_2 - N_1$, of the form

$$N_2 - N_1 = \frac{P_2\tau_2\left(1 - \frac{\tau_1}{\tau_{21}}\right) - P_1\tau_1}{1 + \left(\tau_1 + \tau_2 - \frac{\tau_1\tau_2}{\tau_{21}}\right)\frac{\sigma I}{h\nu}}. \tag{7.45}$$

According to its definition (see equation (7.11*b*)), the *gain* is obtained as $\sigma(N_2 - N_1)$, and can be written in the form

$$\gamma(\nu) = \frac{\gamma_0(\nu)}{1 + I/I_s}, \tag{7.46a}$$

where

$$\gamma_0(\nu) = \sigma(\nu)\left[P_2\tau_2\left(1 - \frac{\tau_1}{\tau_{21}}\right) - P_1\tau_1\right] \tag{7.46b}$$

$$I_s = \frac{h\nu}{\sigma(\nu)\tau_2} \cdot \frac{1}{1 + \frac{\tau_1}{\tau_2}\left(1 - \frac{\tau_2}{\tau_{21}}\right)}. \tag{7.46c}$$

Note that γ_0 has the meaning of a small signal gain, that is $\gamma \to \gamma_0$ for $I \to 0$. The saturation intensity for which the gain drops to $1/2\gamma_0$, now has a more complex form, but reduces to $I_s = \frac{h\nu}{\sigma\tau_2}$(as in equation (7.42)) for $\tau_2 \gg \tau_1$, or $\tau_{21} = \tau_2$.

Other interesting cases are left as exercises. With the Laplace transform presented here, all situations should be mathematically tractable (although some calculations may be lengthy).

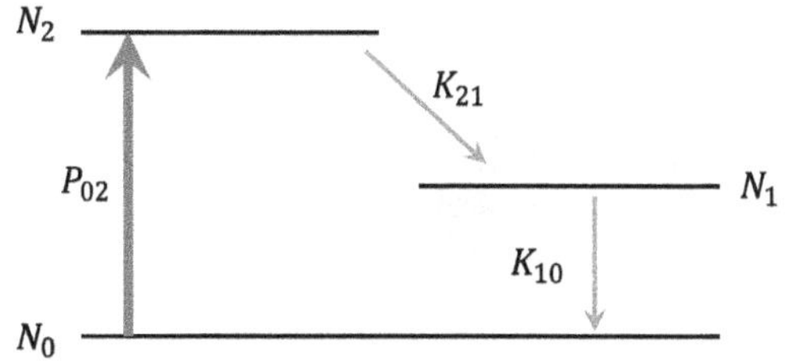

Figure 7.11. Problem 7.1.

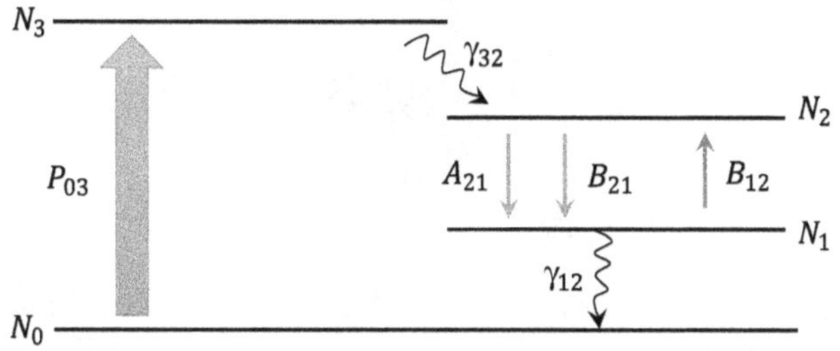

Figure 7.12. Problem 7.2.

7.7 Problems

1. Consider the three-level system depicted in figure 7.11.

   The decay rates indicated represent the inverse lifetimes, for example, $K_{21} = \dfrac{1}{\tau_{21}}$; P_{02} is the pump rate from level 0 to 2. Assume level N_0 does not deplete.

   a) Write the rate equations for the three levels.
   b) Plot each population versus time, for $P_{02} = 10^{23}\,\mathrm{m}^{-3}\,\mathrm{s}^{-1}$, $K_{21} = 10^9$ Hz, $K_{10} = 5 \times 10^9$ Hz. All initial populations are 0 at $t = 0$.
   c) Find the relation between K_{10} and K_{21} for which we obtain a population inversion at $t = t_0$.
   d) Derive the steady state solution for N_0, N_1, N_2.

2. Consider the four-level laser system below (figure 7.12).

   Let $P_{03} = 10^{20}\,\mathrm{m}^{-3}\,\mathrm{s}^{-1}$, $\gamma_{32} = 10^{10}\mathrm{s}^{-1}$ and $\gamma_{12} = 10^9\mathrm{s}^{-1}$ be the two nonradiative decay rates, $A_{21} = 10^9\,\mathrm{s}^{-1}$ the spontaneous emission rate, $B_{21} = 2 \times 10^9\mathrm{s}^{-1}$ the stimulated emission rate, B_{12} the absorption rate, and $B_{12} = B_{21}$. Initially, all concentrations are zero, and the pump is applied at $t = 0$ as the step function,

   $$P_{03}(t) = P_{03}\Gamma(t).$$

   a) Write the rate equations for all levels.
   b) Solve for $N_1(t)$ and $N_2(t)$ and plot $N_2(t) - N_1(t)$. Plot the quantum yield for the transition 2→1 versus time.
   c) Repeat b) for γ_{12} and γ_{32} that are 10× larger and then lower than those given above.
   d) What are the steady state populations under the conditions in a)?

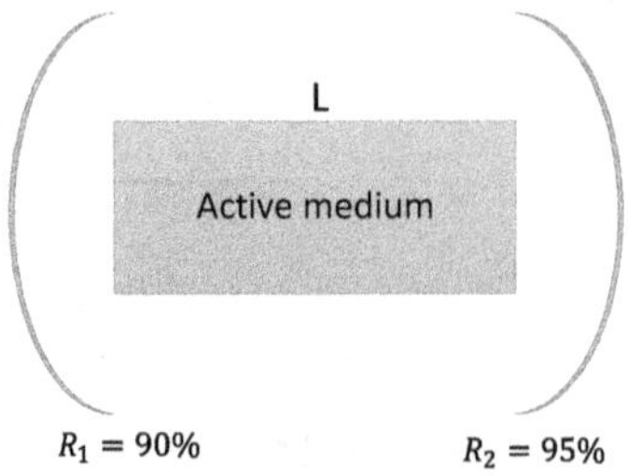

Figure 7.13. Problem 7.5.

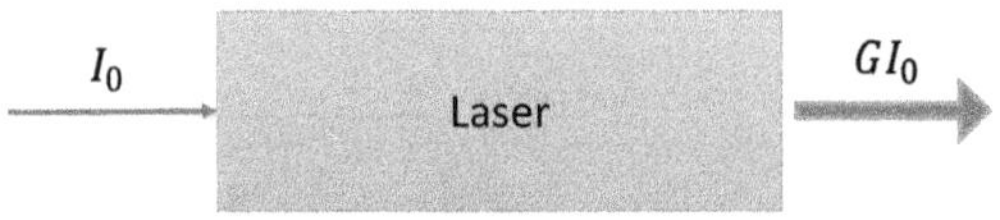

Figure 7.14. Problem 7.9.

3. Prove that laser operation is impossible with a two-level atomic system.
4. The natural bandwidth of a laser is broadened by the Doppler mechanism. However, the gas is flown at a constant speed ν_1 in the laser cavity, say, along the z-axis. The velocity distribution of the gas molecules are described by the Gaussian probability density $p(\nu_z) = \dfrac{1}{\sqrt{2\pi}\,\Delta\nu} e^{-\dfrac{(\nu_z - \nu_1)^2}{2\Delta\nu^2}}$.
   a) Calculate the resulting spectral line, which is now inhomogenously broadened.
   b) Calculate the temporal autocorrelation of the optical field.
5. Calculate the threshold gain coefficient, γ, for the laser active medium that is 1 m long and has reflectivities at the two mirrors of 0.9% and 0.95% (figure 7.13).
6. Solve the problem of the three-level laser system (equation (7.28)), calculate $N_1(t)$ and $N_2(t)$ assuming $1/\tau_2 = 0$ and all the other parameters as non-zero.
7. Same as problem 6, except now $P_1 = 0$ and $\dfrac{1}{\tau_{21}} = 0$.
8. Same as problem 6, except now all parameters are non-zero.
9. A laser amplifier outputs $10I_0$ when input irradiance is $I_0 = 10\ \text{W/cm}^2$ and $7I_0$, when I_0 is $50\ \text{W/cm}^2$ (figure 7.14).
   a) Calculate the saturation intensity.
   b) Calculate the small signal gain.

References

[1] Einstein A 1917 The quantum theory of radiation *Physikalische Zeitschrift* **18** 121

[2] Verdeyen J T 1995 *Laser Electronics* 3rd edn (Prentice Hall Series in Solid State Physical Electronics vol 778) (Englewood Cliffs, NJ: Prentice Hall), xxvi p 12

[3] Arfken G B and Weber H-J 2001 *Mathematical Methods for Physicists* 5th edn (San Diego, CA: Academic), xiv p 1112

IOP Publishing

Principles of Biophotonics, Volume 2
Light emission, detection, and statistics
Gabriel Popescu

Chapter 8

Classification of optical detectors

8.1 Waves and photons

An optical detector converts incident optical radiation into a *measurable* signal, most commonly via an electrical circuit. The discussion here is not limited to the visible portion of the electromagnetic spectrum (see chapter 1), but also covers UV and IR, within the wavelength interval of approximatively 100 nm to 1 mm.

Depending on the physical mechanism that leads to detection, detectors are divided into two classes: *thermal* and *photon*. Thermal detectors respond to a *temperature* change induced by the absorption of the incident radiation. Photon detectors, on the other hand, sense the *direct interaction* between the light and the detector material. The thermal interaction can be described by the classical model of light absorption in the material, meaning the optical radiation can be treated as continuous. However, the operation of photodetectors is best described by using quantified light, that is, *photons*.

We have already seen in chapter 6 how black body radiation was correctly described by Planck [1] using the discrete quanta of energy, $E = h\upsilon$. Later, Einstein used this concept to explain the photoelectric effect [2]. Using this energy to frequency conversion, we established the photon-based radiometric quantities in chapter 3. In photodetectors, the photon energy, $h\upsilon$, is converted into another form of energy, for example, the kinetic energy of an electron, which becomes detectable.

In 1920, de Broglie discovered that all particles behave like waves. Specifically, he showed that a particle of momentum p, has an associated 'de Broglie' wavelength $\lambda = h/p$. Note that, for a non-realistic particle, $\mathbf{p} = m\mathbf{v}$, where m is the mass of the particle and $\mathbf{v}$ the velocity. For photons, the energy and momentum relate to the angular temporal frequency, ω, and the angular spatial frequency (wavevector), $\mathbf{k}$, respectively, as

$$E = \hbar\omega \qquad (8.1a)$$

$$\mathbf{p} = \hbar \mathbf{k}. \quad (8.1b)$$

In chapter 8 of Volume 1, we established the uncertainty relations for classical fields,

$$\Delta\omega\Delta t \geqslant 1/2 \quad (8.2a)$$

$$\Delta k_x \Delta x \geqslant 1/2 \quad (8.2b)$$

where Δ denotes the standard deviation of the intensity distribution in space (x), time (t), spatial frequency (k) and temporal frequency (ω). Combining equations (8.1) and (8.2), we obtain the uncertainty relations for discrete photons,

$$\Delta E \Delta t \geqslant \hbar/2 \quad (8.3a)$$

$$\Delta p_x \Delta x \geqslant \hbar/2. \quad (8.3b)$$

Equation (8.3*a*) establishes that the energy and localization in time of a photon cannot be known with arbitrarily high precision. Similarly, equation (8.3*b*) states that the photon momentum along one axis (x) and its localization along that axis have uncertainties that are inversely proportional. As emphasized before (chapter 8, Volume 1), these uncertainties are not due to our inability to perform measurements. Rather, the uncertainty relation establishes the precision with which the quantities *can be defined* [3]. Thus, measurements cannot provide lower uncertainties than those expressed in equations (8.2) and (8.3).

8.2 Photon detectors

In this volume we discuss three classes of photon detectors capable of directly converting photon fluxes into measurable quantities: *photovoltaic*, *photoconductive*, and *photoemissive* detectors.

Photovoltaic detectors (*photovoltaics*) are solid-state devices capable of producing a voltage or current in the absence of an external bias. Most commonly, a photovoltaic sensor consists of a *semiconductor photodiode*. In chapter 12, we will discuss the properties of semiconductors and the principles of operation for a photodiode. Figure 8.1 describes a photovoltaic device.

Photoconductive devices operate on the principle that light can significantly increase the conductivity of a semiconductor. As a result of the incident light, under constant voltage bias, there will be an increase in the current across the device. This result can be easily understood, since the current density, **j**, is proportional to the conductivity, σ,

$$\mathbf{j} = \sigma \mathbf{E}. \quad (8.4)$$

In equation (8.4), **E** is the electric field applied across the device, $[\mathbf{E}] = V/\mathbf{m}$, $[\mathbf{j}] = A/\mathbf{m}^2$, $[\sigma] = \Omega\ \mathbf{m}$ (ohm meter). In figure 8.2, the electric field across the semiconductor is $E = V/L$, where L is the length of the material. Thus, the current measured by the ammeter is

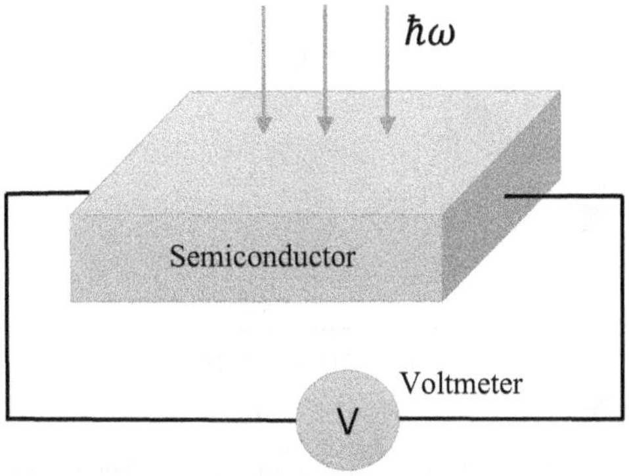

Figure 8.1. Generic representation of a photovoltaic detector: the incident light of frequency ω generates a measurable voltage.

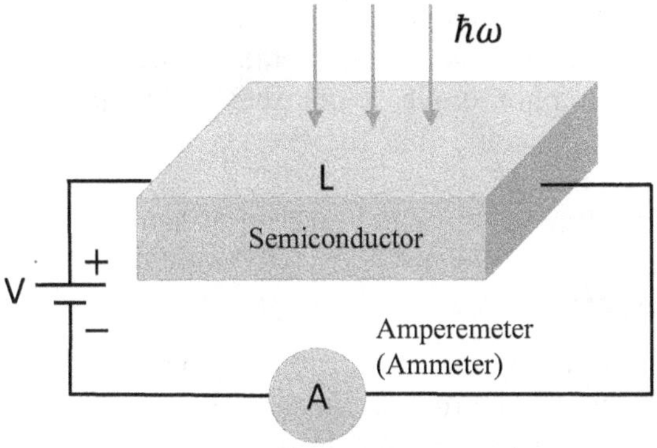

Figure 8.2. Illustration of a photoconductor device. The incident light of frequency ω induces an increase in the material conductivity, which generates a rise in the current measured by the ammeter.

$$
\begin{aligned}
I &= jA \\
&= \sigma E A \\
&= \sigma V \frac{A}{L} \ .
\end{aligned}
\tag{8.5}
$$

Equation (8.5) indicates that the change in current is linear with the change in conductivity. The slope is proportional to the voltage applied and the cross section of the device, but inversely proportional to the length of the device.

Photoemissive detectors rely on the (external) photoelectric effect, which is the emission of electrons from the surface of a material following the absorption of incident photons. Vacuum photodiodes and photomultipliers are common examples of photoemissive detectors. According to the model proposed by Einstein, the kinetic energy, E_c, of the photoelectron extracted via a photon absorption is

$$
E_c = \hbar\omega - W. \tag{8.6}
$$

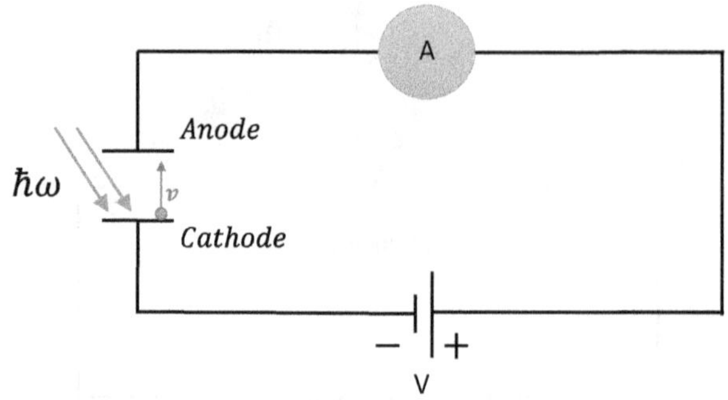

Figure 8.3. Illustration of a vacuum photodiode upon absorption of a photon. An electron leaves the cathode (C) with a velocity v and travels toward the anode (A), thus boosting the current measured by ammeter A.

In equation (8.6), $\hbar\omega$ is the energy of the photon and W is the work necessary to extract the electron from the material (typically, a metal). The energy W is sometimes called the *work function* of the metal. Figure 8.3 illustrates the operating principle of a vacuum photodiode. Note that photoemissive processes have a threshold of operation, meaning that the energy of the photon has to be at least equal to the work function, W.

We will discuss photon detectors in much more detail in chapter 13.

8.3 Thermal detectors

Thermal detectors operate by monitoring the temperature change associated with light absorption. The word 'thermal' originates from the Greek 'thermos', meaning 'heat'. As pointed out by Dereniak and Boreman [4], the fact that IR radiation is sometimes called 'heat radiation' causes some confusion, as IR can also be detected by photon detectors.

The temperature change in the detector produces a corresponding change in properties of the material, as follows.

- *Bolometers* operate by a temperature-induced change in *resistance.*
- *Thermocouples* work by generating a temperature-induced *voltage.*
- *Pyroelectrics* respond to temperature change by a change in *electrical polarization.*

Thermal detectors will be described in more detail in chapter 14.

8.4 Problems

1. Two materials of conductivities σ_1 and σ_2 are subjected to the same voltage V. If the materials have the same cross sections, what is the ratio of the two lengths for which they yield the same current?
2. Photons of wavelength λ generate photoelectrons from a metal of work function W. What is the de Broglie wavelength of the photoelectrons?
3. An optical field of central wavelength λ and bandwidth $\Delta\lambda$ is incident on a photoemissive detector of work function W.

a) What is the temporal uncertainty in localizing these photons?
b) What is the uncertainty in the energy of the emitted photoelectrons?

4. A Gaussian beam of width (standard deviation) Δx is incident on a photoemissive detector. If the work function of the material is W, calculate the smallest angle of divergence that the photoelectrons can exhibit.

References

[1] Planck M and Masius M 1914 *The Theory of Heat Radiation* (Philadelphia, PA: P. Blakiston's Son & Co) xiv p 1 225p

[2] Stachel J J (ed) 1998 *Einstein's Miraculous Year: Five Papers That Changed the Face of Physics* (Princeton, NJ: Princeton University Press)

[3] Heisenberg W 1950 *The Physical Principles of the Quantum Theory* (New York: Dover Publications), 183

[4] Dereniak E L and Boreman G D 1996 *Infrared Detectors and Systems* vol 306 (New York: Wiley)

IOP Publishing

Principles of Biophotonics, Volume 2

Light emission, detection, and statistics

Gabriel Popescu

Chapter 9

Statistics of optical detection

9.1 Probabilities

Photodetection is fundamentally a *stochastic* process. For example, the number of photoelectrons generated by a photoemissive detector within a certain period of time cannot be predicted with perfect accuracy, i.e., it is a *random variable*. Similarly, the noise affecting our measurements is stochastic in nature (see chapter 10 for a description of various types of noise).

In this chapter we establish the basic framework for such a statistical description. In chapter 9 of Volume 1, we discussed random signals and the output they generate through linear systems. Many of those translate to the statistics of detection. Probability is a measure of likelihood for a certain event to occur. Mathematically, the probability $P(E)$ assigns a number within [0, 1] for an event E in a set of possible events (see chapter 4 in [1]). For example, flipping a coin can only yield two possible events: heads or tails. Thus, the probability of getting heads is 1/2. Rolling a die can yield six possible events, thus, the probability of rolling a three, for example, is 1/6.

Let us consider a set of events, D, as shown in figure 9.1. If we consider two subsets of events, E_1, E_2, in general, there might be an overlap between them. For example, if we consider a 52-card deck and extract one card randomly, the subset of events for extracting an *ace* overlaps with the subset of extracting *clubs*. The subset of overlap consists of a single possible event, namely, the ace of clubs.

We denote by $E_1 \cup E_2$ the union of two sets. In the example above, $E_1 \cup E_2$ describes the events of extracting an ace *or* a club (figure 9.1b). $E_1 \cap E_2$ denotes the *intersection* set, the possible events of extracting an ace *and* a club, that is, an ace of clubs (figure 9.1c).

The probability $P(E)$ satisfies the following axioms [1]

$$P(E) \geqslant 0, \tag{9.1}$$

which mean that the probability is non-negative.

doi:10.1088/978-0-7503-1644-6ch9

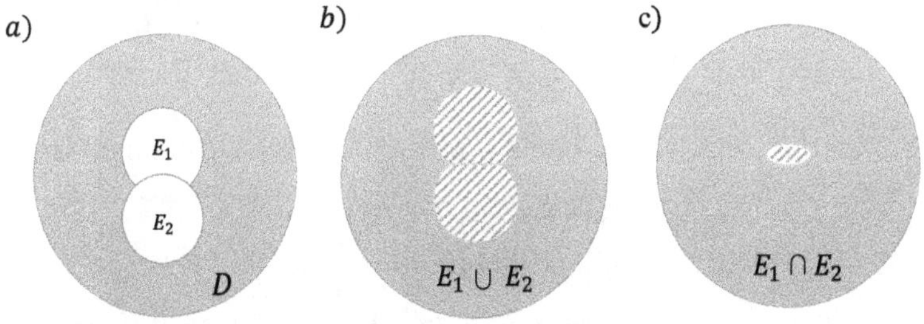

Figure 9.1. (a) Subsets of events E_1, E_2 within the set D. (b) The union set, $E_1 \bigcup E_2$. (c) The intersection set, $E_1 \bigcap E_1$.

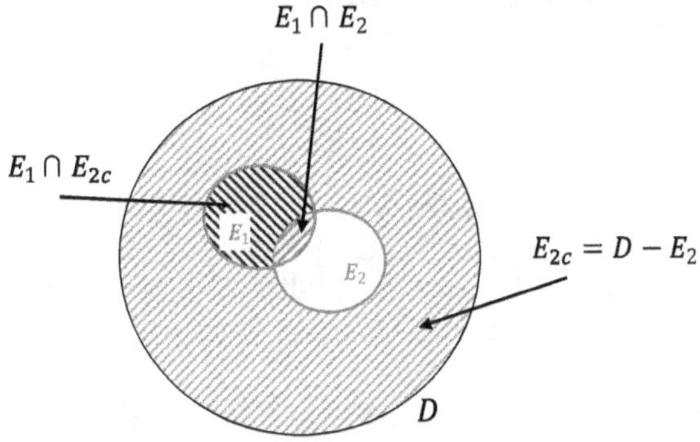

Figure 9.2. Representation of the joint of E_1 and E_{2c}.

$$P(D) = 1, \tag{9.2}$$

meaning that the likelihood of any event in the set D to happen is 100%.

$$P(E_1 \bigcup E_2 \bigcup \ldots \bigcup E_n) = P(E_1) + P(E_2) + \ldots P(E_n), \tag{9.3}$$

for mutually exclusive (independent) events.

From figure 9.1, we see that

$$P(E_1 \bigcup E_2) = P(E_1) + P(E_2) - P(E_1 \bigcap E_2). \tag{9.4}$$

The complement of a set E is denoted by E_c and characterized by the property

$$P(E) + P(E_c) = 1. \tag{9.5}$$

Thus, the complement of the entire event set, D, for which $P(D) = 1$, is

$$P(D_c) = 0. \tag{9.6}$$

For two subsets of D, E_1 and E_2, the following property applies (figure 9.2)

$$\begin{aligned} P(E_1 \bigcap E_{2c}) &= P[E_1 \bigcap (D - E_2)] \\ &= P(E_1 \bigcap D) - P(E_1 \bigcap E_2) \\ &= P(E_1) - P(E_1 \bigcap E_2). \end{aligned} \tag{9.7}$$

A *joint* event, J, is defined as an event that belongs to both E_1 and E_2,

$$J = E_1 \cap E_2. \tag{9.8}$$

The *joint probability*, that is, the likelihood for this event to happen is

$$\begin{aligned} P(J) &= P(E_1 \cap E_2) \\ &= P(E_1E_2). \end{aligned} \tag{9.9}$$

Note that sometimes the simpler notation $P(E_1E_2)$ is used. In the example with the deck of cards, the probability of extracting a club is $13/52 = 1/4$, and the probability of extracting an ace is $4/52 = 1/13$. However, the probability of extracting an ace of clubs is $1/52$ (there is only one such card in the stack). We can see that the joint probability cannot be larger than any of the individual ones. For *independent variables*, the joint probability is

$$P(E_1E_2) = P(E_1)P(E_2).$$

Clearly, $P(ace \cap clubs) = P(clubs)P(ace)$, $1/52 = (1/4)(1/13)$.

The *conditional* probability measures the probability of E_2 to occur, given that E_1 occurred, $P(E_2/E_1)$. In the earlier example: what is the probability of extracting an ace of clubs, $P(ace/clubs)$, given that we know the card is a club, $P(clubs)$? We see that now the probability is higher, $P(ace/clubs) = 1/13$, since there are 13 cards which are clubs, one of which is an ace. Compared to the original, unconditional probability, 1/52, $P(ace/clubs)$ is four times higher. It turns out 1/4 is precisely the probability of drawing from the entire deck a card of clubs. Thus, it can be shown that, for conditional probabilities, the following relation holds

$$P(E_2/E_1) = \frac{P(E_2 \cap E_1)}{P(E_1)}. \tag{9.10}$$

Thus, $P(ace/clubs) = P(ace \cap clubs)/P(clubs)$, or, $1/13 = (1/52)4$.

As a result of equation (9.10), we observe that

$$P(E_2/E_1) \geqslant P(E_2 \cap E_1). \tag{9.11}$$

Importantly, $P(E_2/E_1)$ is generally different from $P(E_1/E_2)$. From equation (9.10), we find that

$$P(E_2/E_1) = P(E_1/E_2) \tag{9.12}$$

only if $P(E_1) = P(E_2)$, since $E_1 \cap E_2 = E_2 \cap E_1$.

9.2 Continuous random variables

So far, we have discussed *discrete* random variables, for which there is a *countable* number of possible events. However, often in practice we deal with *continuous* random variables, whereby each event is surrounded by an infinitesimal region of possible events. Therefore, within a given domain, the total number of possible events is infinite. For example, the distribution of light intensity along a certain

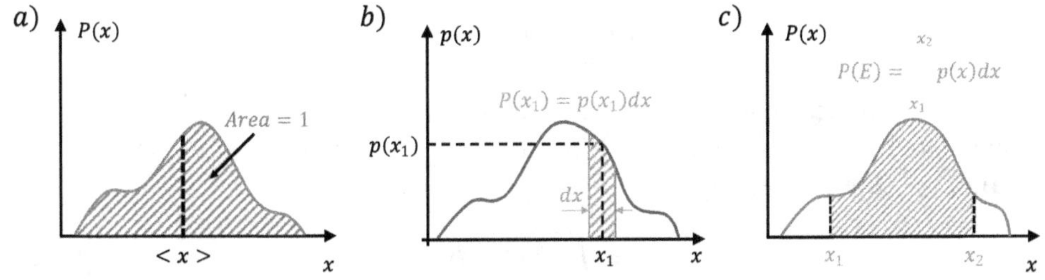

Figure 9.3. (a) The PDF, $p(x)$, is normalized to the unit area. (b) The probability of x taking a value in the vicinity of x_1 equals the infinitesimal area under the curve and around x_1. (c) The probability of x taking a value within the interval (x_1, x_2) equals the area under $p(x)$ over that interval.

direction, $I(x)$, is such a continuous random variable, describing the probability of receiving light in the vicinity of position x. The continuous function that fully describes the statistics of a continuous random variable is called a probability density function (PDF), denoted by $p(x)$, as illustrated in figure 9.3.

The function $p(x)$ is normalized to the unit area (figure 9.3(a))

$$\int_{-\infty}^{\infty} p(x)dx = 1, \tag{9.13}$$

indicating that the probability of having *any* value for x is 100%. The probability of x having a value in the vicinity of x_1 is (figure 9.3(b))

$$p(x_1) = p(x_1)dx. \tag{9.14}$$

Note that the probability of x having the value of precisely x_1 is *zero*, which is the result of the fact that there are an infinite number of possible values for the continuous random variable x. As we narrow the possible value range around x_1, the probability (the area in figure 9.3(b)) vanishes.

The probability for x to take a value within the interval (x_1, x_2) is (figure 9.3(c))

$$P[x \in (x_1, x_2)] = \int_{x_1}^{x_2} p(x)dx. \tag{9.15}$$

9.3 Moments of a distribution

When detecting optical signals, the PDF of interest can be, for example, the distribution of voltage values across a detector. It is very useful to know certain averaged quantities that characterize the distribution: the mean, standard deviation, etc. We have already discussed in Volume 1 the moments of a distribution and even the moments of a convolution (Volume 1, chapters 4–9). Here, we recall briefly the definition of nth order moment

$$<x^n> = \int_{-\infty}^{\infty} x^n p(x) dx = m_n \tag{9.16}$$

where, as usual, $<>$ denotes the (ensemble) average.

Since $p(x)$ is a PDF, that is, it is normalized to the unit area, the average and variance (σ^2) are defined as

$$\begin{aligned} <x> &= \int_{-\infty}^{\infty} x p(x) dx \\ &= m_1 \end{aligned} \tag{9.17a}$$

$$\begin{aligned} \sigma^2 &= <(x - <x>)^2> \\ &= <x^2> - 2 <x <x>> + <x>^2 \\ &= <x^2> - <x>^2 \\ &= m_2 - m_1^2. \end{aligned} \tag{9.17b}$$

The average $<x>$ gives us the expected value of the random variable x and σ informs on how much spread we expect around the main value.

The *characteristic function* associated with a distribution is defined as the expectation value of a complex exponential, of the form $<e^{ikx}>$,

$$\begin{aligned} \psi(k) &= \int_{-\infty}^{\infty} e^{ikx} p(x) dx \\ &= \tilde{p}(k). \end{aligned} \tag{9.18}$$

Equation (9.18) indicates that the characteristic function is nothing more than the Fourier transform of the probability density, $p(x)$.

Let $p(x)$ describe an irradiance distribution of a field along x,

$$p(x) = |U(x)|^2 \tag{9.19a}$$

with

$$\int_{-\infty}^{\infty} |U(x)|^2 \, dx = 1. \tag{9.19b}$$

The characteristic function in this case represents the irradiance-averaged *plane wave*,

$$\begin{aligned} \psi(k) &= <e^{-ikx}>_x \\ &= \int_{-\infty}^{\infty} e^{-ikx} |U(x)|^2 \, dx \\ &= [U \otimes U](k). \end{aligned} \tag{9.20}$$

In equation (9.20), we used the correlation theorem to prove that the characteristic function equals the autocorrelation of the Fourier transform of the field.

Conversely, if the PDF describes the wavevector distribution of the field,

$$p(k) = |U(k)|^2 \tag{9.21a}$$

$$\int_{-\infty}^{\infty} |U(k)|^2\, dk = 1, \tag{9.21b}$$

the wavevector averaged plane wave is

$$\begin{aligned} \psi(x) &= \langle e^{ikx} \rangle_k \\ &= \int_{-\infty}^{\infty} e^{ikx}\, |U(k)|^2\, dk \\ &= [U \otimes U](x). \end{aligned} \tag{9.22}$$

Of course, analog relationships hold for the time-temporal frequency domain

$$\begin{aligned} \psi(\omega) &= \langle e^{i\omega t} \rangle_t \\ &= [U \otimes U](\omega)' \end{aligned} \tag{9.23a}$$

$$\begin{aligned} \psi(t) &= \langle e^{-i\omega t} \rangle_\omega \\ &= [U \otimes U](t). \end{aligned} \tag{9.23b}$$

In equations (9.20), (9.22–9.23), we used the previous sign conventions, whereby $e^{-i(\omega t - kx)}$ is the inverse Fourier transform relationship (see Volume 1, chapter 3)

$$U(x, t) = \int_{-\infty}^{\infty} \int_{-\infty}^{\infty} U(k, \omega) e^{-i(\omega t - kx)} dk\, d\omega. \tag{9.24}$$

9.4 Common probability distributions

9.4.1 Binomial distribution

The *binomial distribution* is a *discrete* probability distribution that describes the number of successes in a number of independent experiments n. The outcome of these experiments allows for only two values: yes (one, true) and no (zero, false). This distribution is useful when describing the probability of detecting photons after a number of experiments. A Bernoulli trial defines a single experiment with either a *true* outcome (of probability p) or a *false* outcome (probability $1 - p$).

The probability of getting k successes after n trials is given by

$$P(k; n; p) = \binom{n}{k} p^k (1 - p)^{n-k}, \tag{9.24a}$$

where,

$$\begin{aligned} \binom{n}{k} &= C_n^k \\ &= \frac{n!}{k!(n - k)!}. \end{aligned} \tag{9.24b}$$

In equation (9.24), C_n^k denotes the number of combinations in which we can make groups of k numbers in a set of n. For example, given a set of four cards, how many groups of two cards can be made? The answer is $C_4^2 = 4!/(2 \cdot 2) = 6$ combinations. C_m^k is sometimes called the binomial coefficient.

The probability function in equation (9.24*a*) has the following interpretation: given n trials, one gets k successes with the probability p^k and $n - k$ failures with the probability $(1 - p)^{n-k}$. However, the k successes can occur in any sequence, such that the total number of combinations in which n trials yields k successes is C_n^k.

It can be easily seen that the binomial distribution is normalized, meaning any number of successes, $k \in [0, n]$, should occur with 100% probability,

$$\begin{aligned}\sum_{k=0}^{n} P(k; n; p) &= \sum_{k=0}^{n} C_n^k p^k (1 - p)^{n-k} \\ &= (p + 1 - p)^m \\ &= 1.\end{aligned} \tag{9.25}$$

In proving equation (9.25), we use the binomial decomposition theorem,

$$(a + b)^m = \sum_{k=0}^{m} C_m^k a^k b^{n-k}. \tag{9.26}$$

The mean of the binomial distribution is

$$\begin{aligned}<k> &= \sum_{k=0}^{n} k C_n^k p^k (1 - p)^{n-k} \\ &= \sum_{k=1}^{n} \frac{n!}{(k-1)!(n-k)!} p^k (1 - p)^{n-k} \\ &= np \sum_{k-1=0}^{n-1} \frac{(n-1)! p^{k-1} (1-p)^{[(n-1)-(k-1)]}}{(k-1)![(h-1)-(k-1)]!} \\ &= np(p + 1 - p)^{n-1} \\ &= np.\end{aligned} \tag{9.27}$$

Equation (9.27) is intuitive: the average number of successes is the number of trials multiplied by the probability of success. For example, rolling a dice 600 times will yield, on average, $600 \times \frac{1}{6} = 100$ sixes.

The variance is calculated according to the definition

$$\sigma^2 = <k^2> - <k>^2 . \tag{9.28}$$

Instead of computing $<k^2>$ directly, it is more practical to calculate $<k(k - 1)>$, because the first two terms in the summation, $k = 0, 1$, vanish

$$
\begin{aligned}
< k(k-1)> &= \sum_{k=2}^{n} k(k-1)\frac{n!}{k!(n-k)!}p^k(1-p)^{n-k} \\
&= n(n-1)^2 p \sum_{k-2=0}^{n-2} \frac{(n-2)!}{(k-2)![(n-2)-(k-2)]!}p^{k-2}(1-p)^{n-k} \\
&= n(n-1)p^2(p+1-p)^{n-2} \\
&= n(n-1)p^2.
\end{aligned}
\tag{9.29}
$$

Since $<k(k-1)> = <k^2> - <k>$, we arrive at the expression for variance as

$$
\begin{aligned}
\sigma^2 &= < k(k-1)> + <k> - <k>^2 \\
&= n(n-1)p^2 + np - (np)^2 \\
&= np[(n-1)p + 1 - np] \\
&= np(1-p).
\end{aligned}
\tag{9.30}
$$

In the earlier example, performing experiments, on an average we obtain 100 sixes, and the standard deviation is $\sigma = \sqrt{600\frac{1}{6}\frac{5}{6}} = 9.1$. Figure 9.4 illustrates examples of the binomial distribution for various values of success probability, p, and number of trails, n.

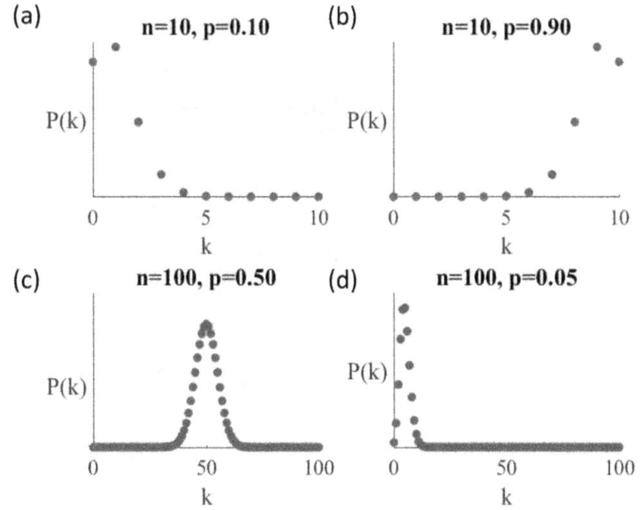

Figure 9.4. Binomial probability density for: (a) $n = 10$, $p = 0.1$; (b) $n = 10$, $p = 0.9$; (c) $n = 100$, $p = 0.5$; (d) $n = 100$, $p = 0.05$.

9.4.2 Poisson distribution

The Poisson distribution is an approximation of the binomial distribution when the number of trials, n, is very large and the probability of success, p, is very low. Let us denote the average number of successes after n trials as $\lambda = np$. We can evaluate the binomial probability density for $p = \lambda/n \to 0$,

$$\begin{aligned} P(k;\lambda) &= \frac{n!}{k!(n-k)!}\left(\frac{\lambda}{n}\right)^k\left(1-\frac{\lambda}{n}\right)^{n-k} \\ &= \frac{\lambda^k}{k!}\frac{n!}{(n-k)!n^k}\frac{\left(1-\frac{\lambda}{n}\right)^n}{\left(1-\frac{\lambda}{n}\right)^k} \\ &= \frac{\lambda^k}{k!}\frac{n(n-1)\ldots(n-k+1)(n-k)!}{n\cdot n\ldots n(n-k)!}\cdot\frac{\left(1-\frac{\lambda}{n}\right)^n}{\left(1-\frac{\lambda}{n}\right)^k}. \end{aligned} \tag{9.31}$$

For $n\to\infty$, we have $\frac{n!}{(n-k)!n^k}\to 1$, $\left(1-\frac{\lambda}{n}\right)^k\to 1$, and $\left(1-\frac{\lambda}{n}\right)^n\to e^{-\lambda}$. The limit can be proven by assuming a continuous function $f(x)=\left(1+\frac{a}{x}\right)^x$ and taking the limit of its logarithm using L'Hopital's theorem,

$$\begin{aligned} \lim_{x\to\infty}\log[f(x)] &= \lim\left[x\log\left(1+\frac{a}{x}\right)\right] \\ &= \lim_{x\to\infty}\frac{\log\left(1+\frac{a}{x}\right)}{\frac{1}{x}} \\ &= \lim_{x\to\infty}\frac{\log\left(1+\frac{a}{x}\right)'}{\left(\frac{1}{x}\right)'} \\ &= \lim_{x\to\infty}\frac{-\frac{a}{x^2}}{1+\frac{a}{x}}\cdot\frac{1}{-\frac{1}{x^2}} \\ &= \lim_{x\to\infty}a\frac{x}{a+x} \\ &= a. \end{aligned}$$

As a result,

$$\lim_{x\to\infty}\left(1+\frac{a}{x}\right)^x = e^a \tag{9.32}$$

and

$$\left(1-\frac{\lambda}{n}\right)^n = e^{-\lambda}. \tag{9.33}$$

Thus, if we plug the results of these limits into expression for the probability, equation (9.31), we obtain

$$P(k,\lambda) = e^{-\lambda}\frac{\lambda^k}{k!}, \tag{9.34}$$

which defines the Poisson distribution. We can think of the Poisson distribution as the binomial distribution in the 'rare event' limit ($p \to 0$).

Figure 9.5 shows a comparison between the Poisson and binomial distributions. It can be seen how the Poisson approximation of the binomial distribution is more accurate as $n \to \infty$ and $p \to 0$.

The expected value for a Poisson distribution is

$$\begin{aligned}<k> &= \sum_{k=0}^{\infty} k e^{-\lambda}\frac{\lambda^k}{k!}\\ &= \lambda.\end{aligned} \tag{9.35}$$

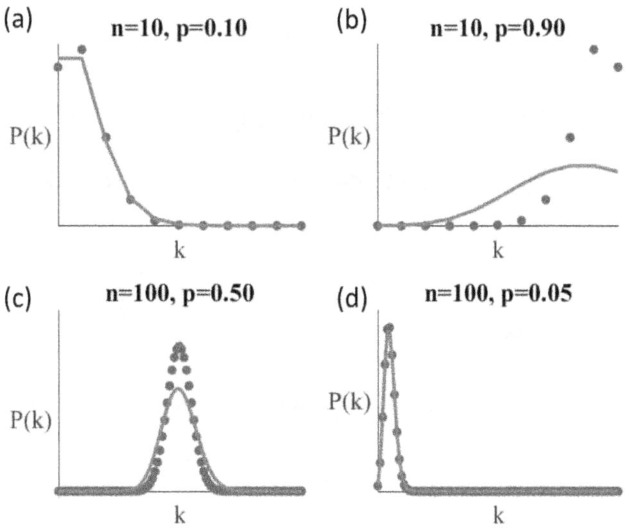

Figure 9.5. The Poisson approximation (solid line) to the binomial probability function (dots): (a) $n = 10$, $p = 0.1$; (b) $n = 10$, $p = 0.9$; (c) $n = 100$, $p = 0.5$; (d) $n = 100$, $p = 0.05$.

The result in equation (9.35) is obtained by recalling that the binomial distribution has a mean $<k> = np = \lambda$, or by using the result of the summation $\sum_{k=0}^{\infty} \lambda^k / k! = \lambda e^{\lambda}$.

The variance is

$$\begin{aligned}\sigma^2 &= \lim_{\substack{n\to\infty \\ p\to 0}} np(1-p) \\ &= np = \lambda.\end{aligned} \tag{9.36}$$

Thus, we find that the variance of a Poisson distribution equals the average, λ.

The Poisson distribution is commonly used to predict the outcome of photon detection under low-flux conditions. The detection of a photon can be represented as a series of delta functions, *randomly* spaced (figure 9.6). Recall from section 3.3 that the photon flux is defined as the *average* number of photons dN delivered per unit time, dt

$$P_q = \frac{dN}{dt}. \tag{9.37}$$

For a *stationary* emission process (see Volume 1, chapter 9), P_q is constant in time. The average number of photons, ΔN, within an interval $\Delta t \to 0$ is

$$\begin{aligned}\Delta N &= P_q \Delta t \\ &= \lambda.\end{aligned} \tag{9.38}$$

At the same time, as $\Delta t \to 0$, the average $\lambda \to 0$, so the probability of detecting a photon within the interval Δt becomes (see equation (9.34))

$$\begin{aligned}p(1, \lambda \to 0) &= e^{-\lambda}\frac{\lambda^1}{1!} \\ &= \lambda \\ &= P_q \Delta t.\end{aligned} \tag{9.39}$$

It follows that the probability to detect zero photons within Δt is

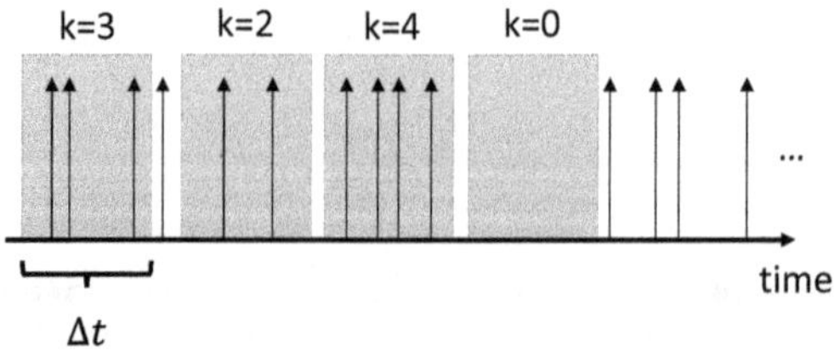

Figure 9.6. Photon detection events represented as series of delta functions randomly spaced. The number of photons detected in a temporal window (acquisition time) of duration Δt is a random variable.

$$P(0, \lambda \to 0) = 1 - \lambda$$
$$= 1 - P_q \Delta t. \qquad (9.40)$$

9.4.3 Gaussian distribution and the central limit theorem

The Gaussian (normal) PDF is defined as (figure 9.7)

$$p(x) = \frac{1}{\sqrt{2\pi}\,\sigma} e^{-\frac{(x-<x>)^2}{2\sigma^2}}. \qquad (9.41)$$

In equation (9.41), $<x>$ and σ define the mean and standard deviation of the distribution. The characteristic function is the Fourier transform of p (see chapter 4 in Volume 1),

$$p(\omega) = e^{-\frac{\sigma^2\omega^2}{2}} e^{i<x>\omega}. \qquad (9.42)$$

The distribution associated with the sum of *independent* random variable approaches a Gaussian as the number of random variables goes to infinity. This statement is known as the *central limit theorem*. In order to gain an intuitive understanding of this fundamental result, let us consider two independent random processes X and Y, characterized by their probability densities, p_X and p_Y, respectively. Let $Z = X + Y$, for which we would like to find the probability density p_Z. In order to find p_Z, we aim to calculate the probability for the *sum* to return the value $Z = z$. If $X = x$, then $Y = z - x$. Thus, the probability to obtain $Z = z$, $p_Z(z)$ is the intersection of the event sets $(X = x)$ and $Y = z - x$, summed for all possible values of x,

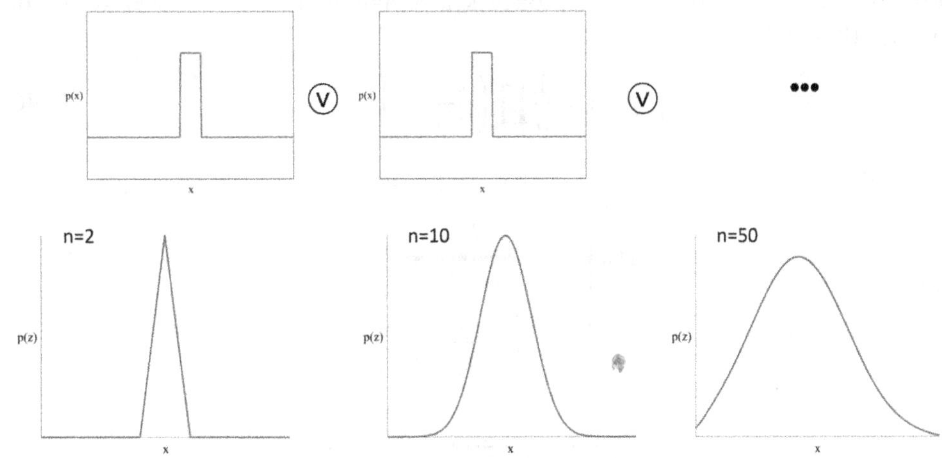

Figure 9.7. Central limit theorem: the PDF for a sum of independent events approaches a Gaussian. Convolving a large number of distributions yields a Gaussian.

$$\begin{aligned} p_Z(z) &= \int_{-\infty}^{\infty} p_X(x) p_Y(z-x) dx \\ &= [p_X \circledcirc p_Y](z). \end{aligned} \tag{9.43}$$

Equation (9.43) states an interesting fact: the PDF of the sum is the convolution of the individual PDFs. It can be shown that, as the number of independent random variables in the sum becomes very large, p_z approaches a Gaussian function (figure 9.7),

$$\begin{aligned} p_z(z) &= [p_{x_1} \circledcirc p_{x_2} \circledcirc \ldots \circledcirc p_{x_n}](z)\Big|_{n\to\infty} \\ &= \frac{1}{\sqrt{2\pi}\,\sigma} e^{\frac{-(z-<z>)^2}{2\sigma^2}}. \end{aligned} \tag{9.44}$$

Using the 'moments of a convolution' property described in Volume 1, section 4.3, we have the following expressions for the mean and variance of $p_z(z)$:

$$<z> = <z>_1 + <z>_2 + \ldots + <z>_n \tag{9.45a}$$

$$\sigma^2 = \sigma_1^2 + \sigma_2^2 + \ldots + \sigma_n^2. \tag{9.45b}$$

Another way of stating the central limit theorem is that a large number of convolution operations converge to a Gaussian shape. One common example of the central limit theorem in action is Brownian motion. A small particle suspended in water at thermal equilibrium is subject to a large number of independent kicks from the water molecules. As a result, the probability of finding the particle at a certain position in space and time approaches a Gaussian.

9.4.4 Uniform distribution

The uniform distribution describes a random process for which all events are equally probable. Thus, the function approaches a rectangular function over a certain interval (figure 9.8)

$$p(x) = \frac{1}{a} \Pi\left(\frac{x - x_0}{a}\right). \tag{9.46}$$

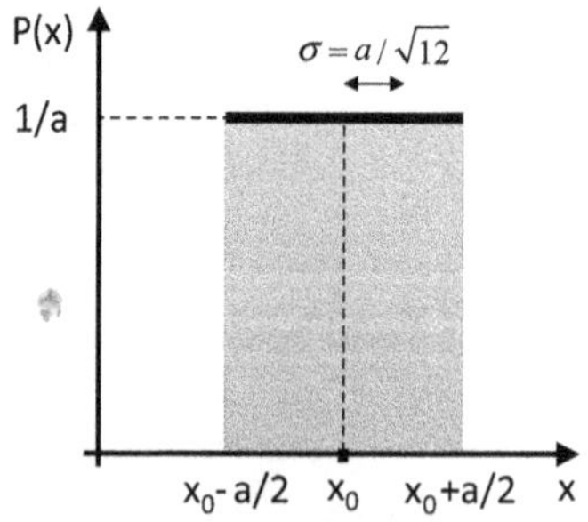

Figure 9.8. The uniform distribution with the average, x_0, and standard deviation, σ, indicated.

The average is

$$
\begin{aligned}
\langle x \rangle &= \int_{-\infty}^{\infty} x p(x) dx \\
&= \frac{1}{a}\int_{-\infty}^{\infty} x \prod\left(\frac{x - x_0}{a}\right) dx \\
&= \frac{1}{a}\int_{-\infty}^{\infty} (x - x_0) \prod\left(\frac{x - x_0}{a}\right) d(x - x_0) \\
&\quad + \frac{1}{a} x_0 \int_{-\infty}^{\infty} \prod\left(\frac{x - x_0}{a}\right) d(x - x_0) \\
&= \frac{1}{a}\int_{-a/2}^{a/2} t\, dt + \frac{1}{a} x_0 \int_{-a/2}^{a/2} dt \\
&= \frac{1}{a}\frac{t^2}{2}\Bigg|_{-a/2}^{a/2} + \frac{x_0}{a} t \Bigg|_{-a/2}^{a/2} \\
&= x_0.
\end{aligned}
\tag{9.47}
$$

Not surprisingly, the average of the uniform distribution is at its center. The second order moment is

$$
\begin{aligned}
\langle x^2 \rangle &= \frac{1}{a}\int_{-\infty}^{\infty} x^2 \left(\frac{x - x_0}{a}\right) dx \\
&= \frac{1}{a}\frac{t^3}{3}\Bigg|_{-a/2}^{a/2} - 2a \langle x \rangle + x_0^2 \\
&= x_0^2 + \frac{a^2}{12}.
\end{aligned}
\tag{9.48}
$$

Thus, the variance of a uniform distribution is

$$
\begin{aligned}
\sigma^2 &= \langle x^2 \rangle - \langle x \rangle^2 \\
&= \frac{a^2}{12}.
\end{aligned}
\tag{9.49}
$$

A common example of the uniform distribution affecting the outcomes of optics experiments is in speckle statistics [2]. Thus, light that scatters a large number of times in an inhomogeneous medium (e.g., tissue or a diffuser) ends up with a distribution of *phase* shifts that is uniformly distributed.

9.4.5 Exponential distribution

The exponential PDF is defined as (figure 9.9)

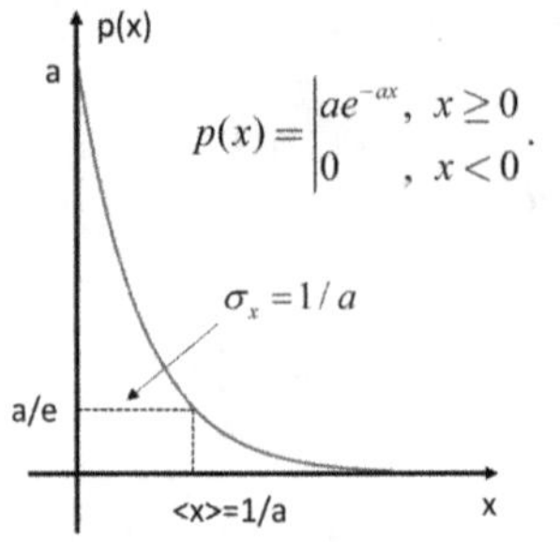

Figure 9.9. Exponential distribution, with mean and standard deviations as indicated.

$$p(x) = \begin{vmatrix} ae^{-ax}, & x \geqslant 0 \\ 0, & x < 0 \end{vmatrix}. \tag{9.50}$$

We can show easily that $\int_{-\infty}^{\infty} p(x)dx = 1$. The average in this case is

$$\begin{aligned} <x> &= \int_{-\infty}^{\infty} xp(x)dx \\ &= \int_0^{\infty} xae^{-ax}\, dx \\ &= -\int_0^{\infty} x(e^{-ax})'dx \\ &= -xe^{-ax}\,|_0^{\infty} + \int_0^{\infty} e^{-ax}\, dx \\ &= -\frac{1}{a}e^{-ax}\Big|_0^{\infty} \\ &= \frac{1}{a}. \end{aligned} \tag{9.51}$$

As a result, the exponential distribution can be expressed as

$$p(x) = \frac{1}{<x>}e^{-\frac{x}{<x>}}. \tag{9.52}$$

The second moment of an exponential distribution is

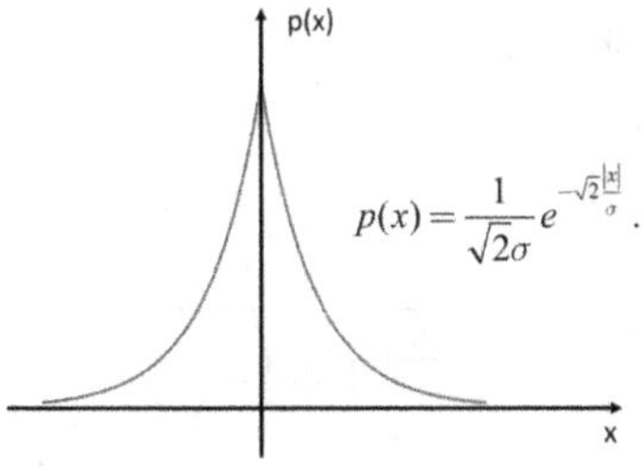

Figure 9.10. Double exponential distribution.

$$
\begin{aligned}
<x^2> &= \int_{-\infty}^{\infty} x^2 p(x)\, dx \\
&= \int_0^{\infty} x^2 a e^{-ax}\, dx \\
&= -\int_0^{\infty} x^2 (e^{-ax})'\, dx \\
&= -x^2 (e^{-ax})|_0^{\infty} + 2\int_0^{\infty} x e^{-ax}\, dx \\
&= \frac{2}{a} <x> \\
&= \frac{2}{a^2}.
\end{aligned} \tag{9.53}
$$

Thus, the variance is

$$
\begin{aligned}
\sigma^2 &= <x^2> - <x>^2 \\
&= \frac{2}{a^2} - \frac{1}{a^2} \\
&= \frac{1}{a^2} \\
&= \langle x \rangle^2.
\end{aligned} \tag{9.54}
$$

9.4.6 Double exponential (Laplacian) distribution

The Laplacian distribution is defined as (figure 9.10)

$$
p(x) = \frac{1}{\sqrt{2}\,\sigma} e^{-\sqrt{2}\frac{|x|}{\sigma}}. \tag{9.55}
$$

Since this distribution is related to the *single-sided* exponential distribution described in section 9.4.5, we leave it as an exercise to calculate $<x>$ and σ.

9.4.7 Lorentzian distribution

The Lorentzian distribution (sometimes known as the Cauchy distribution) is defined as (see figure 9.11(a))

$$p(x) = \frac{1}{\pi \Delta x} \frac{1}{1 + \left(\frac{x}{\Delta x}\right)^2}. \tag{9.56}$$

The moments of a Lorentzian are very particular. While the zeroth order moment (area) equals one, all the higher order moments, $<x>$, $<x^2>$, ... diverge. We can see this peculiar behavior immediately by recalling the moment theorem from Volume 1, section 4.3,

$$<x^n> = i^n \frac{\partial \tilde{f}(k)}{\partial k^n}\bigg|_{k=0}. \tag{9.57}$$

Equation (9.57) (see also equation (4.37) in Volume 1) indicates that the nth order moment can be expressed in terms of the Fourier transform, $\tilde{f}$, differentiated at the origin. We recall that the Fourier transform of a Lorentzian is the double exponential function (figure 9.11(b)):

$$\tilde{f}(k) = e^{-\Delta x|k|}. \tag{9.58}$$

We observe immediately that the first order derivative (slope) of $\tilde{f}(k)$ is not continuous at the origin,

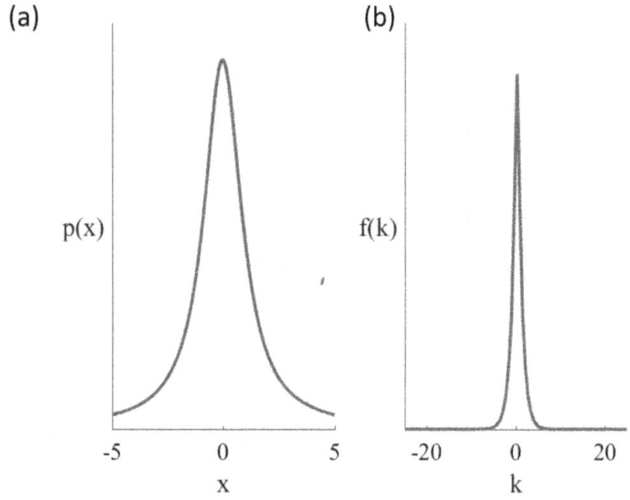

Figure 9.11. (a) Lorentzian distribution. (b) Double exponential function, the Fourier transform of the function in (a).

$$\left.\frac{\partial \tilde{f}(k)}{\partial k}\right|_{k=0} = -\Delta x e^{-\Delta x|k|} \left.\frac{\partial |k|}{\partial k}\right|_{k=0} = \left| \begin{array}{l} +\Delta x,\ k = 0_+ \\ -\Delta x,\ k = 0_- \end{array} \right| . \tag{9.59}$$

Thus, $\partial \tilde{f} / \partial k$ is not continuous at $k = 0$, and therefore it is not derivable at the origin, which means that $<x>$ does not converge. Following the same procedure, one can show that $<x^2>$ also diverges (see problem 9.13). The divergence of the high order moments can be attributed to the slowly varying tails that the Lorentzian exhibits. For example, a Gaussian drops much faster at large x-values, which ensures that its higher moments converge.

9.5 Problems

1. Calculate the probability of rolling two sixes in a row with a die.
2. What is the probability of extracting at random a royal flush from a deck of cards?
3. What is the probability of extracting a royal flush in the order: ten, jack, queen, king, ace, of the same suit?
4. The PDF for a certain random process is

$$p(x) = a_1 e^{-\frac{(x-x_1)^2}{2\Delta x_1^2}} + a_2 e^{-\frac{(x-x_2)^2}{2\Delta x_2^2}},$$

where

$$\int_{-\infty}^{\infty} p(x) dx = 1.$$

   a) Calculate the average, $<x>$.
   b) Calculate the standard deviation, σ_x.
   c) Calculate the nth–order moment of the distribution.
   d) What is the probability of having values in the interval (x_1, x_2)?
5. Calculate the characteristic function, $p(k)$, associated with the distribution $p(x)$ in problem 4. Express the results in problem 4, a), b), c), in terms of $p(k)$.
6. Calculate the following parameters associated with the distribution $|p(k)|^2$ from problem 5:
   a) $<k>$
   b) $<k^2>$.
7. Calculate the third-order moment of the binomial distribution.
8. A photon flux generates photoelectrons with 10% quantum efficiency (10% of photons are converted into electrons). If the incident power is $P = 1$ W at a wavelength $\lambda = 1$ μm, what is the standard deviation of the number of generated photoelectrons during an acquisition time of 1 ms ?

9. A 1 μW power optical field of wavelength $\lambda = 0.5\ \mu m$ is incident on a photodetector. What is the probability of finding zero photons within an interval $\Delta t = 1\ \mu s$?
10. The distribution of phase values in a certain optical field is uniform over the interval $(-\pi, \pi)$.
    a) What is the probability of finding values in the interval $(0, \pi/4)$?
    b) What is the standard deviation of the phase?
11. A PDF is a summation of two single-sided exponentials

$$p(x) = \begin{cases} a_1 e^{-\frac{x}{b_1}} + a_2 e^{-\frac{x}{b_2}}, & x \geqslant 0 \\ 0, & \text{rest} \end{cases}$$

where

$$\int_{-\infty}^{\infty} p(x)dx = 1.$$

Calculate:
    a) $<x>$
    b) $<x^2>$
    c) σ_x (standard deviation).
12. Calculate the characteristic function, $p(k)$, of the distribution in problem 11.
13. Prove that all the moments of a Lorentzian, except for the zeroth-order, diverge.

References

[1] Dereniak E L and Boreman G D 1996 *Infrared Detectors and Systems* vol 306 (New York: Wiley)

[2] Goodman J W 2007 *Speckle Phenomena in Optics: Theory and Applications* (Englewood, CO: Roberts & Co), xvi p 387

IOP Publishing

Principles of Biophotonics, Volume 2
Light emission, detection, and statistics
Gabriel Popescu

Chapter 10

Detection noise

10.1 Mechanisms of noise generation

An optical detector can be described as a system that receives light as *input*, for instance photon flux, P_q, and outputs an electrical quantity such as voltage, ν, or current, i (see figure 10.1).

We can describe the input–output relationship via the operator L that describes the detector:

$$L(P_q) = \nu, \tag{10.1}$$

where we consider that the output signal is a voltage.

Noise represents the random fluctuation in the output ν. Figure 10.2 illustrates a signal outputted by an ideal detector (zero noise) and a real one. Looking at equation (10.1), we see that the random fluctuation in ν can be due to fluctuations in the input, P_q, or the system's response, L, (or both)

$$\delta\nu = \delta P_q + \delta L \tag{10.2}$$

where δ denotes random fluctuations.

There is an intrinsic randomness associated with the input photon flux, δP_q. Thus, the number of photons that reach the detector in a given interval of time is a random variable, characterized by the Poisson statistics (recall section 9.4 and figure 9.6). This contribution to the fluctuations in the output is known as *shot noise* and will be discussed later (section 10.3.2). The noise generated by the system breaks down into noise produced by the detector itself, and noise originating in the electronics accompanying the detector (figure 10.3).

Local fluctuations of the carrier concentration in a resistor can occur due to thermal noise. This contribution to the output noise is called *Johnson noise,* and will be discussed in section 10.3.1. Generation–recombination noise refers to the fluctuations in the rate of generation and recombination of the carriers in a detector

doi:10.1088/978-0-7503-1644-6ch10

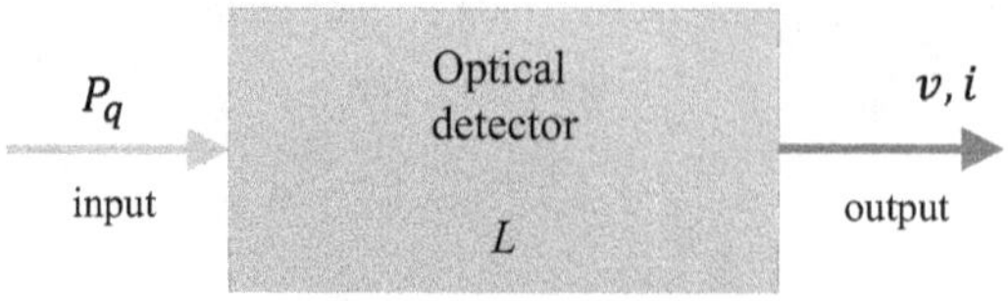

Figure 10.1. An optical detector as a system, of operator L, which converts a photon flux (P_g) into a voltage (ν) or current (i).

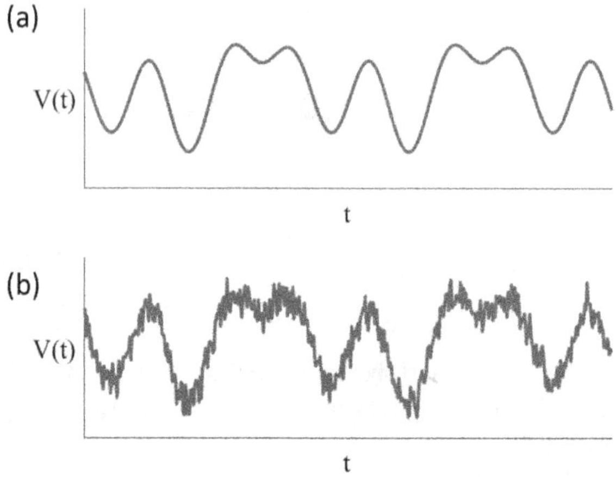

Figure 10.2. Voltage in an ideal (a) and real (b) detector.

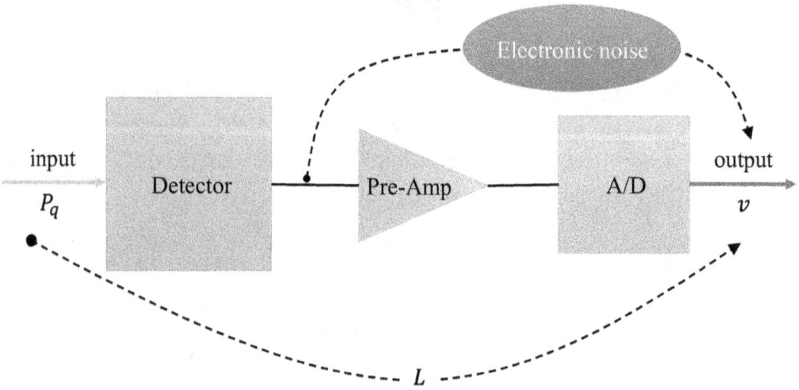

Figure 10.3. The system noise consists of *detector* and *electronic* noise. The electronic noise originates in the pre-amplifier and analog-to-digital (A/D) converter. L is the operator that transforms the input photon flux into an output voltage.

material, and is discussed later in section 10.3.3. One-over-f ($1/f$) noise is a class of noise that becomes significant at low frequencies, f (when $1/f$ is large). This type of noise is discussed in more detail in section 10.3.4. Electronic noise includes all contribution due to the electronic interface that relays the detector signal to the user.

In converting an analog signal into a digital one, *rounding* and *truncation* noise may occur (see section 10.3.5).

Two-dimensional detectors exhibit fluctuations in their output in both time and space. Next, we discuss the basic statistical description of noise as a spatio–temporal random variable.

10.2 Spatio–temporal noise description

10.2.1 Temporal noise

Let us consider the output voltage of a point detector, or a pixel in a 2D detector array, $\nu(t)$. As discussed in Volume 1, chapter 9, the random signal $\nu(t)$ might not have a Fourier transform. For example, if $\nu(t)$ is stationary, it is not square-integrable, and thus its Fourier transform does not exist. Such signals are described using the generalized harmonic analysis, as introduced by Wiener [1] (see also [2]). In order to assign a power spectrum to the random signal $\nu(t)$, we consider first its *truncated* version, denoted by $\underline{\nu}$,

$$\underline{\nu}(t) = \begin{vmatrix} \nu(t), & t \in \left(-\frac{T}{2}, \frac{T}{2}\right). \\ 0, & \text{rest} \end{vmatrix} \tag{10.3}$$

Because of its finite support, $\nu(t)$ does have a Fourier transform, defined as

$$\begin{aligned} \underline{\nu}(\omega) &= \int_{-\infty}^{\infty} \underline{\nu}(t)\, e^{i\omega t}\, dt \\ &= \int_{-\frac{T}{2}}^{\frac{T}{2}} \nu(t)\, e^{i\omega t}\, dt. \end{aligned} \tag{10.4}$$

As usual, we use the same symbol for the signal and its Fourier transform, but carry the argument, such that there is no confusion that, for example, $\nu(t) \leftrightarrow \nu(\omega)$.

The power spectrum associated with $\underline{\nu}$ is

$$\underline{S}(\omega) = <|\underline{\nu}(\omega)|^2> , \tag{10.5}$$

where <> denotes ensemble averaging. In practice, this ensemble average can be performed by measuring sequences of $\underline{\nu}(t)$, of equal duration, computing the Fourier transform of each sequence, and averaging their square magnitudes. The power spectrum of the original signal, $\nu(t)$, is defined in the limit of large T,

$$S(\omega) = \lim_{T\to\infty} \underline{S}(\omega). \tag{10.6}$$

For stationary signals, the autocorrelation function is defined as

$$\Gamma(\tau) = <\nu(t)\, \nu(t+\tau)> , \tag{10.7}$$

where <> denotes, again, the ensemble average. For *ergodic* signals, we can use the ensemble and time average interchangeably (see Volume 1, chapter 9). Thus, $\Gamma(\tau)$ can be expressed in terms of the truncated signal $\underline{\nu}$, as

$$\begin{aligned}\Gamma(\tau) &= \lim_{T\to\infty} < \underline{\nu}(t)\underline{\nu}(t+\tau)> \\ &= \lim_{T\to\infty}\left\{\frac{1}{T}\int_{-T/2}^{T/2} \nu(t)\nu(t+\tau)dt\right\}.\end{aligned} \tag{10.8}$$

The Wiener–Khintchin theorem states that Γ and S form a Fourier transform pair (see chapter 9 in Volume 1)

$$S(\omega) \leftrightarrow \Gamma(\tau)\ . \tag{10.9}$$

We recall that Parseval's theorem is expressed as

$$\int_{-\infty}^{\infty}\Gamma(\tau)d\tau = \frac{1}{2\pi}\int_{-\infty}^{\infty} S(\omega)\, d\omega, \tag{10.10}$$

which states that the total power should be conserved, whether the signal is represented in time or frequency.

The *variance* of the signal is defined as

$$\begin{aligned}\sigma_\nu^2 &= <(\nu - <\nu>^2)> \\ &= \lim_{T\to\infty}\frac{1}{T}\int_{-T/2}^{T/2}(\nu - <\nu>)^2\, dt \\ &= <\nu^2> - <\nu>^2\ .\end{aligned} \tag{10.11}$$

Figure 10.4 provides a comparison between $<\nu^2>$ computed as the time and ensemble average. The random variable $\nu(t)$ is characterized by a probability density $P(\nu)$, which is the histogram of the values, normalized to the unit area, $\int P(\nu)\, d\nu = 1$ (figure 10.4(d)). The ensemble-averaged mean-square is

$$<\nu^2> = \int_{-\infty}^{\infty}\nu^2 P(\nu)\, d\nu, \tag{10.12}$$

which is in the area of overlap in figure 10.4(e). The root mean square, $\sqrt{<\nu^2>}$, (variance for zero-averaged signals) is the side of the square of the same area. For ergodic signals, the two averages described in figure 10.4 are equal.

Combining equations (10.8), (10.10) and (10.11), we see that

$$\begin{aligned}<\nu^2> &= \Gamma(\tau = 0) \\ &= \frac{1}{2\pi}\int_{-\infty}^{\infty} S(\omega)\, d\omega,\end{aligned} \tag{10.13}$$

where the second line is an expression of the central ordinate theorem (Volume 1, chapter 4). Next, we can show that, for stationary signals, $<\nu^2(\tau)> = \Gamma(0) \geqslant |\Gamma(\tau)|$. Consider the inequality

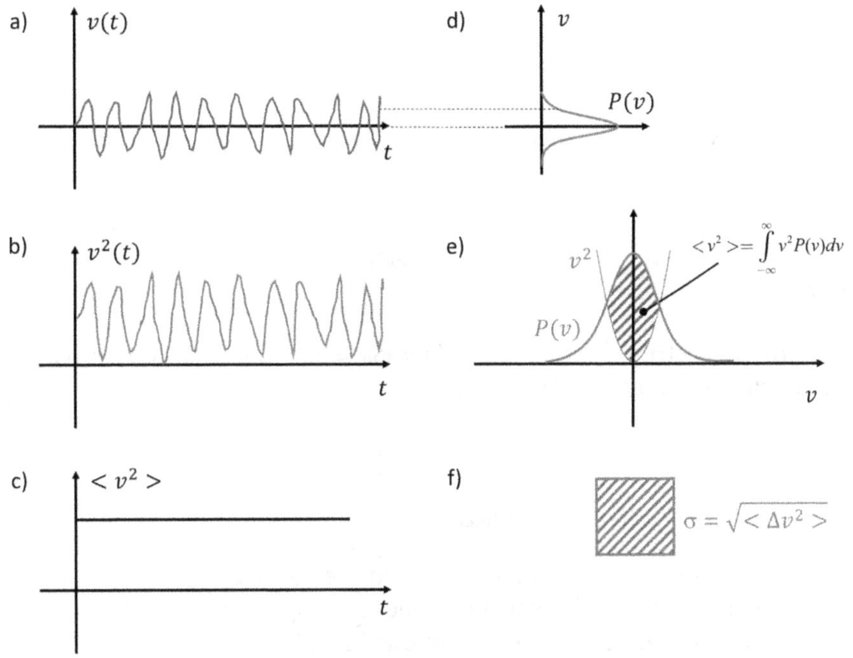

Figure 10.4. (a)–(c) Time-average procedure for calculating $<\nu^2>$ for a zero-averaged signal. (d) The probability density is obtained by projecting the ν-values along the time axis, i.e., plotting the histogram, and normalizing it to the unit area. (e) Graphic representation of $<\nu^2>$ as the area under the product of $P(\nu)$ and ν^2. (f) The standard deviation can be visualized as the side of a square whose area is $<\nu^2>$.

$$[\nu(t) \pm \nu(t+\tau)]^2 = \nu^2(t) + \nu^2(t+\tau) \pm 2\nu(t)\nu(t+\tau) \geqslant 0. \tag{10.14}$$

Taking the time average, we obtain

$$\langle \nu^2 \rangle \geqslant \pm\Gamma(\tau), \tag{10.15}$$

or

$$\langle \nu^2 \rangle \geqslant |\Gamma(\tau)|. \tag{10.16}$$

Thus, we can conclude that the autocorrelation of a stationary signal attains its maximum at the origin, $\Gamma(0) \geqslant |\Gamma(\tau)|$.

If the noise is zero-average, then equation (10.13) indicates that the integral over all frequencies of the power spectrum equals the variance of the signal (see figure 10.5).

$$\int_{-\infty}^{\infty} S(f) df = \sigma_\nu^2, \tag{10.17}$$

where $f = \omega/2\pi$. In describing detector performance, the frequency f is used more often than the angular frequency ω.

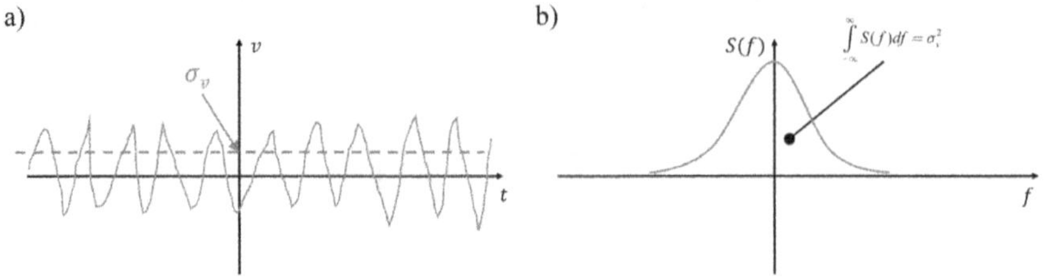

Figure 10.5. (a) Zero-averaged noise signal, $v(t)$, of standard deviation σ. (b) The area under the power spectrum equals the variance.

If the power spectrum is uniform, $S(f) = S_0 = \text{const}$, over a certain bandwidth, Δf, that is, in conditions of *white noise*,

$$\begin{aligned}\sigma_\nu^2 &= \int_{-\Delta f/2}^{\Delta f/2} S_0\, df \\ &= S_0 \Delta f.\end{aligned} \tag{10.18}$$

In other words, equation (10.18) indicates that the variance of the noise is proportional to the bandwidth. Equivalently, the standard deviation of the noise is proportional to the square root of the bandwidth,

$$\sigma_\nu \alpha \sqrt{\Delta f}. \tag{10.19}$$

If the noise signal is generated by a number of *independent* signals, we can show that the total variance is the sum of the individual variances. Thus, using equation (10.13), we obtain

$$\begin{aligned}\sigma^2 &= \Gamma(\tau = 0) \\ &= \left[\sum_{i=1}^{N} \nu_i(t)\right] \otimes \left[\sum_{j=1}^{N} \nu_j(t)\right](\tau)\Bigg|_{\tau=0}.\end{aligned} \tag{10.20}$$

As usual, $\otimes$ in equation (10.20) stands for the correlation operator. The autocorrelation operation in equation (10.20) contains *autocorrelation* terms $\nu_i \otimes \nu_j|_{i=j}$, and *cross-correlation* terms, $\nu_i \otimes \nu_j|_{i\neq j}$, namely

$$\sigma^2 = \left[\sum_{i=1}^{N} \nu_i(t) \otimes \nu_i(t)\right](\tau)\Bigg|_{\tau=0} + \sum_{\substack{i=1\\ i\neq j}}^{N}\sum_{j=1}^{N} [\nu_i(t) \otimes \nu_j(t)](\tau)\Bigg|_{\tau=0}. \tag{10.21}$$

The assumption of *independent* noise sources implies that the cross-correlation terms vanish. As a result, we arrive at the anticipated result that the total variance is the sum of individual variances (figure 10.6)

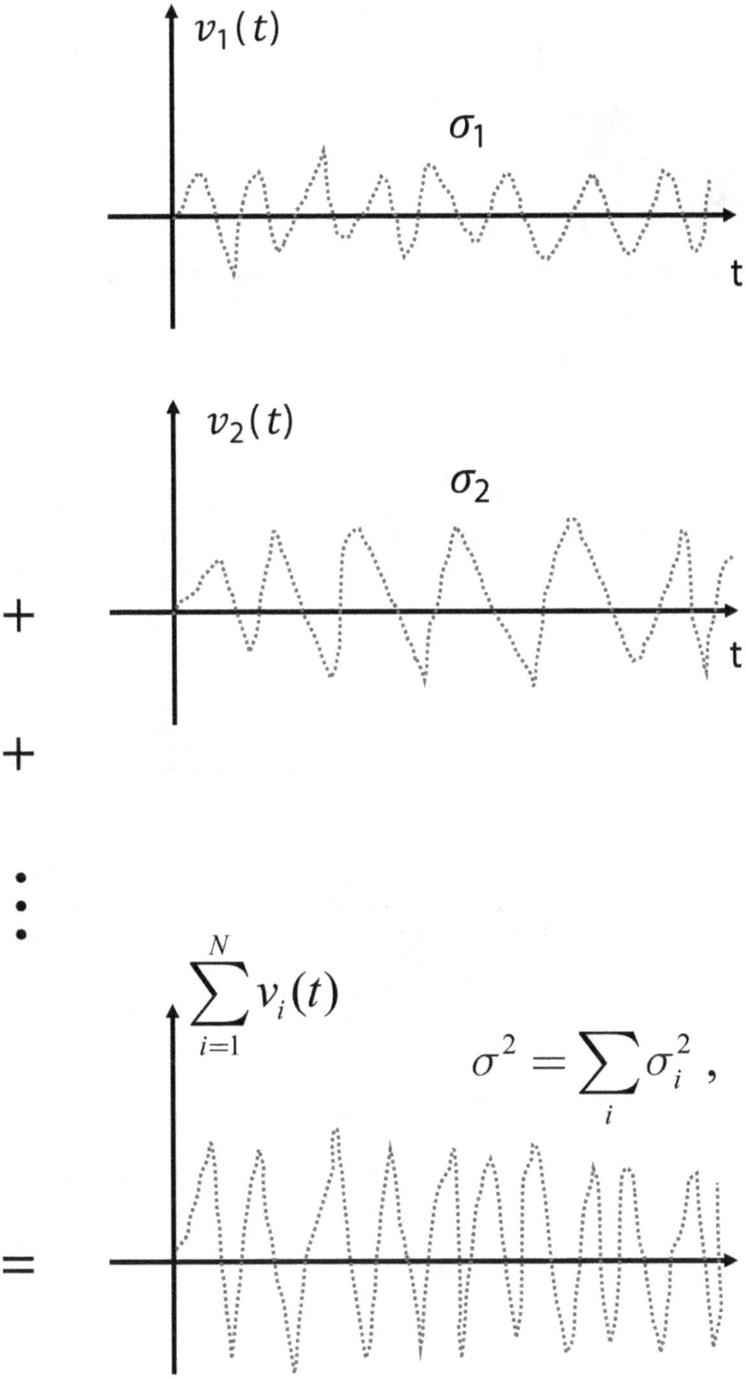

Figure 10.6. Variance of a sum of independent noise signals is the sum of the individual variances.

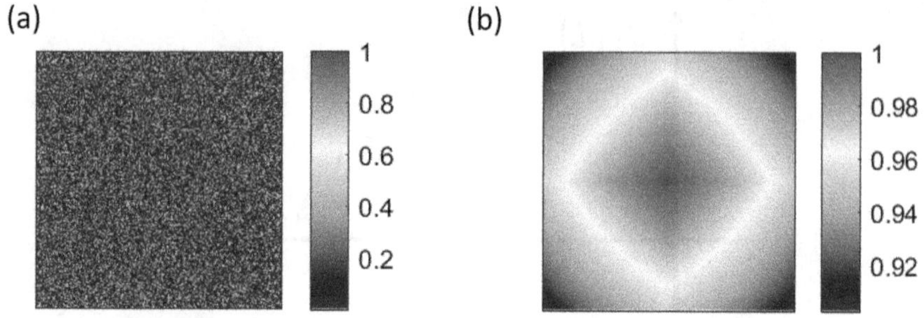

Figure 10.7. a) Spatial noise distribution from a detector array. (b) Spatial autocorrelation function of the noise in (a).

$$\sigma^2 = \sum_i \sigma_i^2, \tag{10.22}$$

As discussed in section 9.4.3, for a large number of independent noise sources, we can predict that the PDF approaches a Gaussian (central limit theorem). This is the underlying reason why Gaussian noise distributions are often used in the literature.

10.2.2 Spatial noise

An entirely analogous description can be performed for spatially fluctuating noise, $\nu(x, y)$. Such a description can be useful when dealing with 2D array detectors, such as CCDs and CMOS. Under the *ergodicity* assumption, we can write the spatial autocorrelation function as (figure 10.7)

$$W(\boldsymbol{\rho}_\perp) = <\nu(\mathbf{r}_\perp)\nu(\mathbf{r}_\perp + \boldsymbol{\rho}_\perp)> , \tag{10.23}$$

where $\mathbf{r}_\perp = (x, y)$ and $\boldsymbol{\rho}_\perp$ is a 2D coordinate shift. The 2D spatial power spectrum is the Fourier transform of W, according to the Wiener–Khintchin theorem,

$$S(\mathbf{k}_\perp) \leftrightarrow W(\boldsymbol{\rho}_\perp). \tag{10.24}$$

The spatial variance of the noise is

$$\sigma^2 = \frac{1}{A}\int_A [\nu(\mathbf{r}_\perp) - <\nu(\mathbf{r}_\perp) >]^2 \, d^2\mathbf{r}_\perp, \tag{10.25}$$

where A is the area of interest (e.g., the detector's surface). The mean-square value is the autocorrelation at the origin, as in the time domain,

$$\begin{aligned} <\nu^2> &= W(\boldsymbol{\rho}_\perp = 0) \\ &= \frac{1}{(2\pi)^2}\int_{-\infty}^{\infty}\int_{-\infty}^{\infty} S(\mathbf{k}_\perp)\, d^2\mathbf{k}_\perp. \end{aligned} \tag{10.26}$$

If we deal with white noise (spatially), $S(\mathbf{k}_\perp) = S_0$, over a certain bandwidth,

$$\Delta\mathbf{k}_\perp = \Delta k_x \Delta k_y,$$

$$
\begin{aligned}
<\nu^2> &= S_0 \frac{\Delta k_x}{2\pi} \frac{\Delta k_j}{2\pi} \\
&= S_0 \Delta f_x \Delta f_y,
\end{aligned}
\tag{10.27}
$$

where f_x, f_y are spatial frequencies in units of m^{-1}. For pixelated detectors with pixel dimensions of $\delta x \times \delta y$, the maximum bandwidths can be expressed as

$$
\Delta f_x = \frac{1}{\delta x} \tag{10.28a}
$$

$$
\Delta f_y = \frac{1}{\delta y}. \tag{10.28b}
$$

Thus, sampling the signal with a pixel size δx can capture the maximum frequency, $f_x^{\max} = \frac{1}{2\delta x}$, according to the Nyquist theorem. The spread between this frequency and its negative counterpart, $-1/2\delta x$, gives us the bandwidth.

If the spatial noise has multiple, independent sources, the overall variance of the signal is the sum of the individual variances,

$$
\sigma^2 = \sum_{i=1}^{N} \sigma_i^2. \tag{10.29}
$$

10.2.3 Averaging

If we deal with noise that fluctuates both spatially and temporally, $\nu(\mathbf{r}_\perp, t)$, we can introduce a spatiotemporal correlation function, $\wedge$, and a spatiotemporal power spectrum, S

$$
\wedge(\boldsymbol{\rho}_\perp, \tau) = <\nu(\mathbf{r}_\perp, t)\nu(\mathbf{r}_\perp + \boldsymbol{\rho}_\perp, t + \tau)>_{\mathbf{r}_\perp, t} \tag{10.30a}
$$

$$
S(\mathbf{k}_\perp, \omega) = \int_{-\infty}^{\infty}\int_{-\infty}^{\infty}\int_{-\infty}^{\infty} \wedge(\boldsymbol{\rho}_\perp, \tau) e^{-i(\omega\tau - \mathbf{k}_\perp \cdot \boldsymbol{\rho}_\perp)} d^2\boldsymbol{\rho}_\perp d\tau. \tag{10.30b}
$$

We can define the spatiotemporal variance, σ^2, as

$$
\sigma^2 = \frac{1}{(2\pi)^3} \int_{-\infty}^{\infty}\int_{-\infty}^{\infty}\int_{-\infty}^{\infty} S(\mathbf{k}_\perp, \omega)\, d^2\mathbf{k}_\perp d\omega, \tag{10.31}
$$

where we assume zero-average noise, $\sigma^2 = <\nu^2>$.

Note that if we fix the temporal frequency $\omega = \omega_0$, the signal becomes deterministic in space, meaning that we can describe the spatial fluctuations decoupled from the temporal ones. Thus, the spatial correlation function at fixed temporal frequency, sometimes referred to as *cross-spectral density*, is defined as

$$
W(\boldsymbol{\rho}_\perp, \omega_0) \int_{-\infty}^{\infty} \wedge(\boldsymbol{\rho}_\perp, \tau) e^{i\omega_0\tau}\, d\tau. \tag{10.32}
$$

In general, the spatial autocorrelation W will depend on w_0. Conversely, we can describe the temporal fluctuations at a given spatial frequency, $\mathbf{k}_\perp = \mathbf{k}_\perp^0$. Thus, the temporal autocorrelation fluctuation at $\mathbf{k}_\perp^0$ is defined as

$$\Gamma(\mathbf{k}_\perp^0, \tau) = \int_{-\infty}^{\infty}\int_{-\infty}^{\infty} \Lambda(\boldsymbol{\rho}_\perp, \tau)e^{-i\mathbf{k}_\perp^0 \cdot \boldsymbol{\rho}_\perp}\, d^2\boldsymbol{\rho}_\perp. \tag{10.33}$$

The temporal autocorrelation function, Γ, generally varies with spatial frequency, $\mathbf{k}_\perp^0$.

Averaging a number of noise measurements results in lowering the mean-square voltage $<\upsilon^2>$. We can express the average noise as

$$\upsilon_N = \frac{1}{N}\sum_{i=1}^{N}\upsilon_i, \tag{10.34}$$

which can vary in time, space, or both, where υ_i is an individual measurement, and N denotes the number of measurements (or *realizations*). If the noise is *uncorrelated*, meaning that each measurement, υ_i, is independent of the others, then the cross-correlation, $\upsilon_i \otimes \upsilon_j = 0$, for $i \neq j$ ($\upsilon_i \otimes \upsilon_j$ operates in space, time, or both). As a result, we can invoke the result in equation (10.22), by which the variance of the sum is the sum of variances. Furthermore, since the noise is assumed *stationary*, all υ_i signals are characterized by the same variance, σ^2. Finally, combining equations (10.22) and (10.34), we obtain the variance of the averaged noise

$$\begin{aligned}\sigma_N^2 &= N < \frac{\upsilon_i^2}{N^2} > \\ &= \frac{\sigma^2}{N}.\end{aligned} \tag{10.35}$$

Equation (10.35) points to a practical method for lowering the noise: averaging N measurements lowers the standard deviation by $\sqrt{N}$. Note that this result is only valid because of our assumption of *uncorrelated stationary* noise across different measurements. If this assumption is not valid, then the averaging becomes less effective. Mathematically, the cross-correlation terms in equation (10.21) contribute to the sum. In the extreme case of *fully correlated* noise across the measurements, the averaging fails entirely to lower σ^2. In this case, from equation (10.24), we see that the variance of the sum is N^2 times the individual variances, meaning that equation (10.35) now yields $\sigma_N^2 = \sigma_i^2$. A comparison between correlated and uncorrelated noise is shown in figure 10.8.

Since the variance is the integral of the power spectrum, let us investigate the effect of averaging in the frequency domain. We start by considering a long (temporal or spatial) sequence of noise, which we then break into shorter sequences, to be averaged (figure 10.9). In order to gain an understanding of the effects that the interval of observation, say duration τ, has on the frequency content of the signal, let us compare the signal, υ_T, of duration T, and a portion of it, n_i, of duration T/N. We consider that they are both portions of an infinite sequence, υ, thus

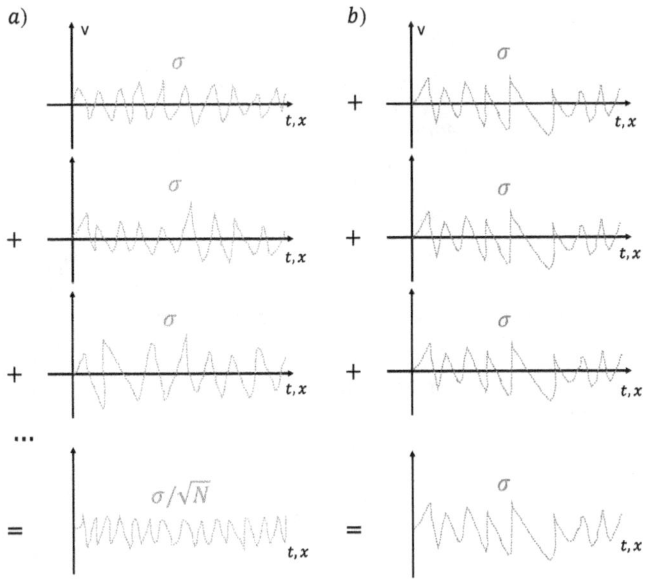

Figure 10.8. (a) Averaging uncorrelated noise reduces σ by $\sqrt{N}$. (b) Averaging fully correlated noise leaves σ unchanged. Noise fluctuates in time (t) or space (x).

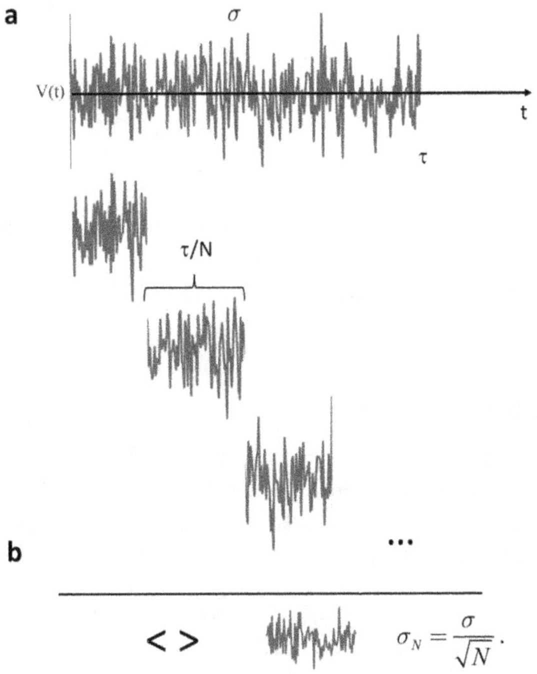

Figure 10.9. (a) Taking a sequence of duration τ and changing it into shorter sequences, of duration $\frac{T}{N}$. (b) Averaging the shorter sequences reduces the standard deviation by $\sqrt{N}$.

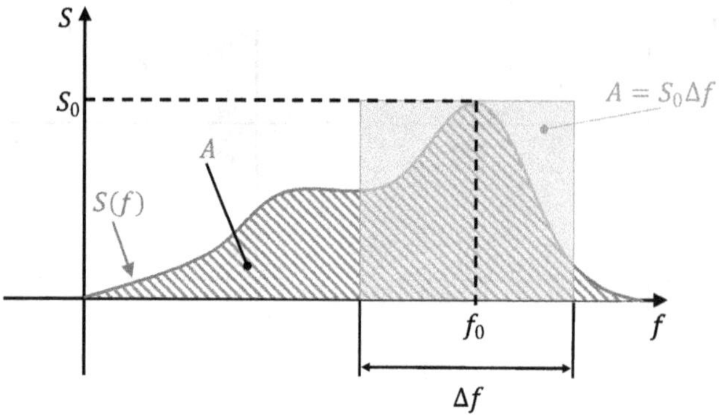

Figure 10.10. Equivalent noise bandwidth, Δf, is the width of the rectangular spectrum of the same area as $S(f)$. S_0 is the maximum value of $S(f)$.

$$\nu_T(t) = \nu(t) \prod\left(\frac{t}{T}\right) \tag{10.36a}$$

$$\nu_i(t) = \nu(t) \prod\left(\frac{t}{T/N}\right). \tag{10.36b}$$

As usual $\prod$ denotes our usual rectangular function. The Fourier transform of these signals have the form

$$\nu_T(\omega) = T\nu(\omega) \circledcirc sinc\ (\omega T) \tag{10.37a}$$

$$\nu_i(\omega) = \frac{T}{N}\nu(\omega) \circledcirc sinc\left(\frac{\omega T}{N}\right). \tag{10.37b}$$

We see from equations (10.37*a*–*b*), that shortening the interval from T to T/N introduces a broader *sinc* function, of width N/T. In other words, shortening the observation time has a smoothing effect on the spectrum, via the convolution operation, $\circledcirc$.

10.2.4 Noise-equivalent bandwidth

Defining the width of a noise power spectrum is subject to a convention, for example, full (half) width half (full) maximum, width at $1/e$, or $1/e^2$, etc. The standard deviation, σ, of the power spectrum is perhaps the most physical, as it also relates directly to the uncertainty relation (see Volume 1, chapter 8),

$$\sigma^2 = \frac{\int_0^\infty (f - <f>)^2 S(f)\, df}{\int_0^\infty S(f)\, df}. \tag{10.38}$$

However, for an arbitrarily shaped power spectrum, computing σ^2 might not be straightforward (see figure 10.10). Often, a simpler definition for the noise bandwidth is used instead: the width of a flat spectrum (i.e., rectangular function) that has a height equal to the maximum of the original spectrum and width such that the area equals that of the original spectrum (figure 10.10).

Let f_0 be the frequency for which $S(f)$ is maximum, $S(f_0) = S_0$. The noise-equivalent bandwidth is defined as (see also Volume 1, chapter 4, problem 23)

$$\Delta f = \frac{\int_0^{\infty} S(f)\, df}{S_0}. \tag{10.39}$$

If voltage ν is the noise signal, we recall that the integral of the spectrum yields the autocorrelation at the origin, which is just the mean squared voltage (see section 10.2.1).

$$\begin{aligned} <\nu(t)\nu(t+\tau)>|_{\tau=0} &= <\nu^2> \\ &= \int_0^{\infty} S(f)\, df. \end{aligned} \tag{10.40}$$

As the average noise power is proportional to $<\nu^2>$, it becomes apparent that Δf is defined such that the power is conserved (hence the name 'equivalent power'). Of course, the analog definition holds for the noise-equivalent spatial bandwidth.

10.3 Noise contributions

10.3.1 Johnson noise

In 1928, Johnson discovered that thermal fluctuations of electrons within a resistor give rise to fluctuations in the voltage across the resistor [3]. This type of thermal noise is known as *Johnson noise.* In order to estimate the mean square voltage fluctuations at thermal equilibrium, we consider the Johnson noise equivalent circuit in figure 10.11. At thermal equilibrium, the equipartition theorem requires that the average energy stored in the capacitor, $\frac{1}{2}C\overline{\nu^2}$, is equal to $\frac{1}{2}k_BT$, where C is the capacitance and T is the absolute temperature. As a result, the mean-squared voltage is

$$\overline{\nu^2} = \frac{k_BT}{C}. \tag{10.41}$$

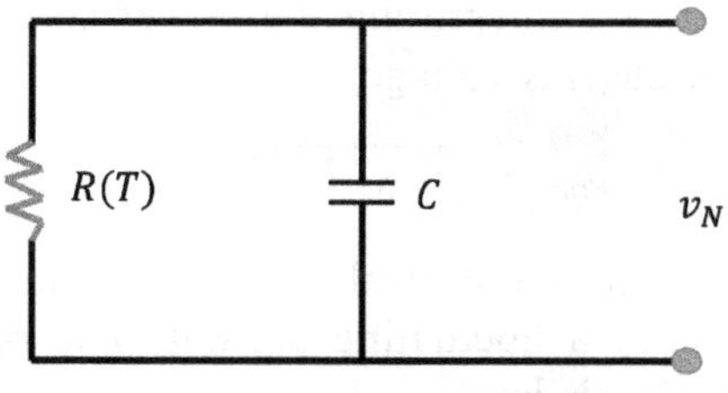

Figure 10.11. Electric circuit for calculating the Johnson voltage noise.

Next, let us describe the temporal autocorrelation of the voltage fluctuations and the power spectrum. Due to the time-constant associated with the RC circuit in figure 10.11, the voltage at the moment t, $\nu(t)$, relates to the voltage at time $t + \tau$, $\nu(t + \tau)$ via

$$\nu(t + \tau) = \nu(t)e^{-\frac{\tau}{RC}}. \tag{10.42}$$

Multiplying equation (10.42) on both sides by $\nu(t)$ and averaging with respect to t, we obtain the voltage autocorrelation as

$$\begin{aligned}\Gamma_\nu(\tau) &= <\nu(t)\nu(t+\tau)>_t \\ &= <\nu^2(t)>_t \, e^{-\frac{|\tau|}{RC}} \\ &= \overline{\nu^2}e^{-\frac{|\tau|}{RC}}.\end{aligned} \tag{10.43}$$

Using the Wiener–Khintchin theorem, we obtain the power spectrum of the Johnson noise as the Fourier transform of Γ (see Volume 1, table 4.1, pp 4–31),

$$S_\nu(f) = 2\overline{\nu^2}\frac{2RC}{1 + (2\pi f RC)^2}. \tag{10.44}$$

In equation (10.44), we used the Fourier property of a double exponential, which gives a Lorentzian function, $e^{-\alpha|t|} \leftrightarrow \frac{2\alpha}{\alpha^2 + \omega^2}$. The additional factor of two in equation (10.44) is due to the fact that $S_\nu(f)$ is integrated over the positive frequency only,

$$\int_0^\infty S_\nu(f)\, df = \overline{\nu^2}, \tag{10.45}$$

which means that S_ν is a factor of two larger than the typical power spectrum, defined over $(-\infty, \infty)$. For frequencies $f \ll \frac{1}{RC}$, equation (10.44) simplifies to

$$S_\nu(f) = 4\overline{\nu^2}RC. \tag{10.46}$$

Using equation (10.41) to express $\overline{\nu^2}C$ in terms of k_BT, we finally obtain the Johnson noise power spectrum

$$S_\nu(f) = 4k_BTR. \tag{10.47}$$

Equation (10.47) is known as the Nyquist formula [4]. Under the low frequency approximation, the Lorentzian spectrum approaches a uniform distribution around the origin (figure 10.12). Under these circumstances, Johnson noise is just a particular type of *white noise.* Integrating the power spectrum over a bandwidth of interest, Δf, we obtain the *rms* voltage,

$$\nu_{rms} = \sqrt{4k_BTR\Delta f}. \tag{10.48}$$

Equation (10.48) represents the central result for Johnson noise. A fluctuating voltage, of course, generates a fluctuating current. The *rms* current, i_{rms}, can be obtained from ν_{rms} using Ohm's law,

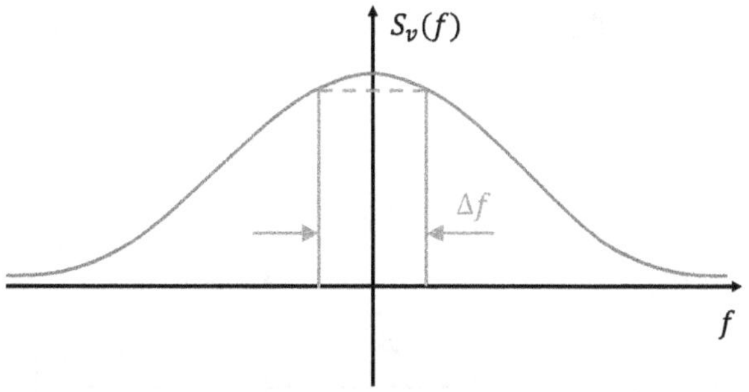

Figure 10.12. Johnson noise spectrum approximated by a uniform distribution at low frequencies.

$$\begin{aligned} i_{rms} &= \frac{v_{rms}}{R} \\ &= \frac{\sqrt{4k_BT\Delta f}}{R}. \end{aligned} \tag{10.49}$$

10.3.2 Shot noise

Shot noise was first studied by Schottky in 1918 [5]. The shot noise contribution is because both the light and the generated photocurrent are discrete in nature. The photon flux and electron current are defined by how many photons or electrons, respectively, are delivered per unit of time. For high photon fluxes, when the number of photons per acquisition time is large, fluctuations of the number of photons around the mean is small. However, in the limit of low photon flux, or short acquisition times, the number of photons detected per each interval can be significant. This *photon noise* generates random fluctuation in the photocurrent, that is, *shot noise*. This low photon flux or *shot noise* limited regime is sometimes called the quantum noise regime. As such, often shot noise represents the lowest noise limit one can achieve with a photon detector. Note that thermal detectors do not exhibit shot noise, as in this case the discrete nature of light is averaged out by the thermal process.

Let us consider the statistics of photon arrival within a certain time interval, T. The probability to detect N photons per time interval T is described by the Poisson statistics (see section 9.4.2)

$$p(N) = \frac{\overline{N}^N}{N!} e^{-N}. \tag{10.50}$$

In equation (10.50), $\overline{N}$ is the *average* number of photons reaching the detector during duration T. Figure 10.13 illustrates the photon detection as a Poisson process.

We found in section 9.4.2 that the variance of a Poisson process equals its mean,

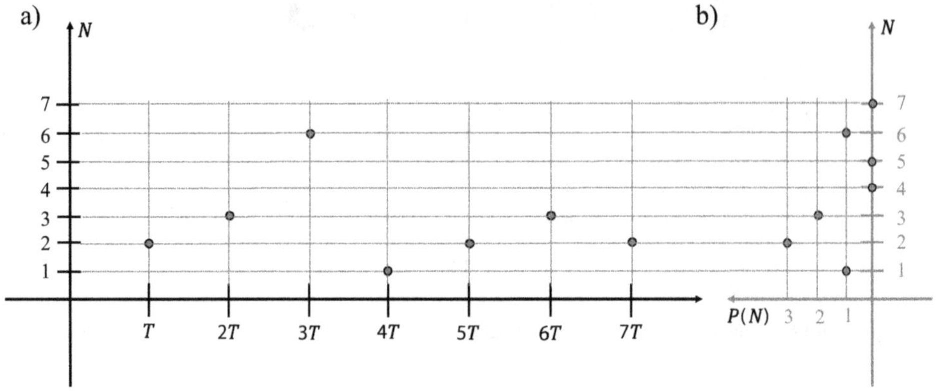

Figure 10.13. (a) The number of photons, N, arriving at the detector during intervals of duration T is a random variable. (b) Probability density, $p(N)$, is a normalized histogram.

$$\begin{aligned}\sigma_N^2 &= <(N - \overline{N})^2> \\ &= \overline{N}.\end{aligned} \tag{10.51}$$

Thus, we can calculate the variance in the photocurrent as a quantitative measure of shot noise. The *average* photocurrent is, by definition, the average charge transported per unit of time,

$$\bar{i} = \frac{\overline{N}e}{T}, \tag{10.52}$$

where e is the elementary charge. The variance of i is

$$\begin{aligned}\sigma_i^2 &= \frac{e^2}{T^2}\sigma_N^2 \\ &= \frac{e^2}{T^2}\overline{N} \\ &= \frac{e}{T}\bar{i}.\end{aligned} \tag{10.53}$$

We can convert the acquisition time, T, into bandwidth, $\Delta f = 1/(2T)$. Thus, we can express the standard deviation of the photocurrent as

$$\sigma_i = \sqrt{2e\bar{i}\Delta f}. \tag{10.54}$$

Equation (10.54) shows that the standard deviation of the noise increases with both the average photocurrent and bandwidth, but only as a square root, that is, a mild dependence. The signal-to-noise ratio (SNR) is

$$\begin{aligned}\mathrm{SNR} &= \frac{\bar{i}}{\sigma_i} \\ &= \sqrt{\frac{\bar{i}}{2e\Delta f}} \\ &= \sqrt{\overline{\mathbf{N}}}.\end{aligned} \tag{10.55}$$

Clearly, the SNR increases with the square root of the number of photons detected. This is not surprising, as we stated initially that the quantum fluctuations in the photon flux are significant only at low signals.

In typical experiments, the input optical signal is described by the power, P, in Watts (section 2.3), or photon flux, P_q, in s^{-1} (see section 3.3). The average current, $\bar{i}$, can be obtained from these quantities as

$$\begin{aligned}\bar{i} &= \eta e P_q \\ &= \eta e \frac{P}{h\upsilon}.\end{aligned} \tag{10.56}$$

In equation (10.56), η is the quantum efficiency, the ratio between the number of photoelectrons and that of incident photons ($\eta \leqslant 1$), h is Planck's constant, and υ is the optical frequency.

Finally, we can express the SNR in terms of these measurable quantities. By combining equations (10.55) and (10.56), we obtain

$$\begin{aligned}\mathrm{SNR} &= \sqrt{\frac{\eta P_q}{2\Delta f}} \\ &= \sqrt{\frac{\eta P}{2h\upsilon\Delta f}}.\end{aligned} \tag{10.57}$$

10.3.3 Generation–recombination noise

Generation–recombination noise rises from the fluctuations in the rate of generation and recombination of carriers [6]. Upon absorbing a photon, an electron receives the excess energy, and becomes more mobile, transitioning from the valence to the conduction band. However, the electron will recombine randomly and return to the valence band, on average after a certain characteristic time, τ_0, called the lifetime. Since both the generation and recombination process are random, the number of electrons in the conduction band, generating the photocurrent in the detector, also fluctuates. The statistics of this generation–recombination process follows the Poisson statistics, such that the variance equals the mean

$$\sigma_N^2 = \overline{N}. \tag{10.58}$$

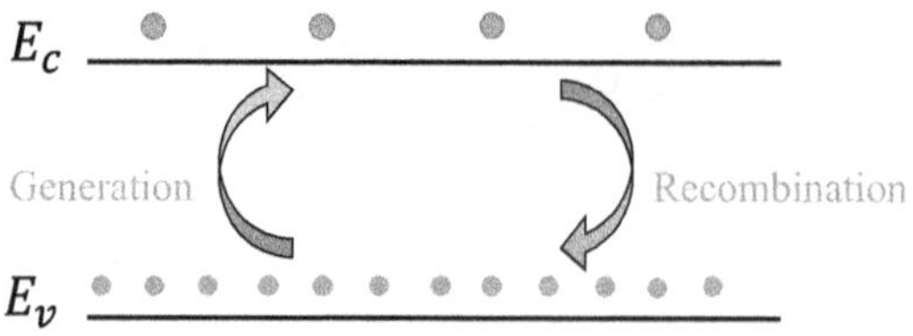

Figure 10.14. Generation–recombination process in a semiconductor detector.

The generation of carriers to the conduction band and the recombination to the valence band are analog to a two-level atomic system encountered when discussing fluorescence (see section 5.1 and figure 10.14). As in fluorescence, the temporal autocorrelation of the number of electrons in the conduction band has an exponential behavior:

$$\begin{aligned}\Gamma_{\Delta N}(\tau) &= \langle\Delta N(t)\Delta N(t+\tau)\rangle_t \\ &= \sigma_N^2 e^{-\frac{\tau}{\tau_0}}.\end{aligned} \tag{10.59}$$

In equation (10.59), $\Delta N = N - \overline{N}$ is the fluctuation around the average of the photoelectron number and $\sigma_N^2 = \langle\Delta N^2\rangle$ is the variance. The power spectrum of these fluctuations is just the Fourier transform of $\Gamma_{\Delta N}$ (see Volume 1, chapter 4, table 4.1 and equation (10.44))

$$S_{\Delta N}(f) = \frac{4\sigma_N^2\tau_0}{1+(2\pi f\tau_0)^2}. \tag{10.60}$$

Note that $S(f)$ is defined as $S_{\Delta N}(f) = 2\int_0^\infty \Gamma_{\Delta N}(\tau)e^{i2\pi f\tau}\,d\tau$, which explains the additional factor of two (as in equation (10.44), equation (10.60) contains a factor of four instead of the usual two). For frequencies much smaller than the inverse lifetime, $f \ll 1/\tau_0$, equation (10.60) simplifies to

$$S_{\Delta N}(f) = 4\overline{N}\tau_0 \tag{10.61}$$

where we used that $\sigma_{\Delta N}^2 = \overline{N}$ for Poisson processes.

To find the current fluctuations, we note that the fluctuation relative to the mean in the electron number and current are proportional to each other [7],

$$\frac{\Delta i}{\overline{i}} = 4G\frac{\Delta N}{\overline{N}}. \tag{10.62}$$

In equation (10.62), the proportionality constant G is called the *photoconductive gain*. The factor G is defined as the ratio between the lifetime τ_0 and the transit time τ_1, which is the time it takes for the electron, on average, to travel between the two electrodes, $G = \tau_0/\tau_1$.

Squaring both sides of equation (10.62) and averaging, we obtain

$$\langle\Delta i^2\rangle = 4G^2\sigma_N^2\frac{\overline{i}^2}{\overline{N}^2} = G^2\frac{\overline{i}^2}{\overline{N}}. \tag{10.63}$$

Since $\bar{i} = \eta e P_q \Delta f$,

$$<\Delta i^2> = 4e^2G^2[\eta P_q \Delta f]. \tag{10.64}$$

Finally, the rms current noise is

$$\Delta i_{rms} = 2eG\sqrt{\eta P_q \Delta f}. \tag{10.65a}$$

Sometimes, the *rms* current is expressed in terms of the photon irradiance, $I_q = \frac{dP_q}{dA}$ (see section 3.6.),

$$\Delta i_{rms} = 2eG\sqrt{\eta I_q A_d \Delta f}, \tag{10.65b}$$

where now A_d is the detector area.

10.3.4 1/f noise

'One-over-f' noise refers to a contribution characterized by a power spectrum of the form

$$S(f) = \frac{A}{f^{\alpha}}, \tag{10.66}$$

where, typically, $\alpha \simeq 1$. Clearly, this type of noise, which affects both thermal and photon detectors, becomes significant at low frequencies.

Compared to *white noise*, where $S(f) = \text{const.}$, $1/f$ noise has higher contributions from low frequencies, which is the reason why it is sometimes called *pink noise* ($1/f$ noise across the visible spectrum would appear pink, due to bias toward the red color). $1/f$ noise is often the reason why we perform measurements at higher frequency by modulating the signal of interest accordingly.

10.3.5 Electronic noise

All optical detectors must be converted to an electronic interface, which itself can introduce noise (see section 10.1). Converting an *analog* (continuous) signal into a *digital* (discrete) one introduces an inherent approximation and, thus, noise.

Let us consider an analog signal $\nu \in [0, \ \nu_{\max}]$. Typically, the quantification level is expressed in the number of bits, b, such that the signal range is sampled in the interval (see figure 10.15)

$$\Delta \nu = \frac{\nu_{\max}}{2^b}. \tag{10.67}$$

Clearly, as $\Delta \nu \to 0$, the resulting digital signal approaches the analog one with no errors. In order to calculate this *rounding noise*, let us consider that the values of ν are uniformly distributed over the range $\Delta \nu$ (see figure 10.15(c)).

The normalized probability density is

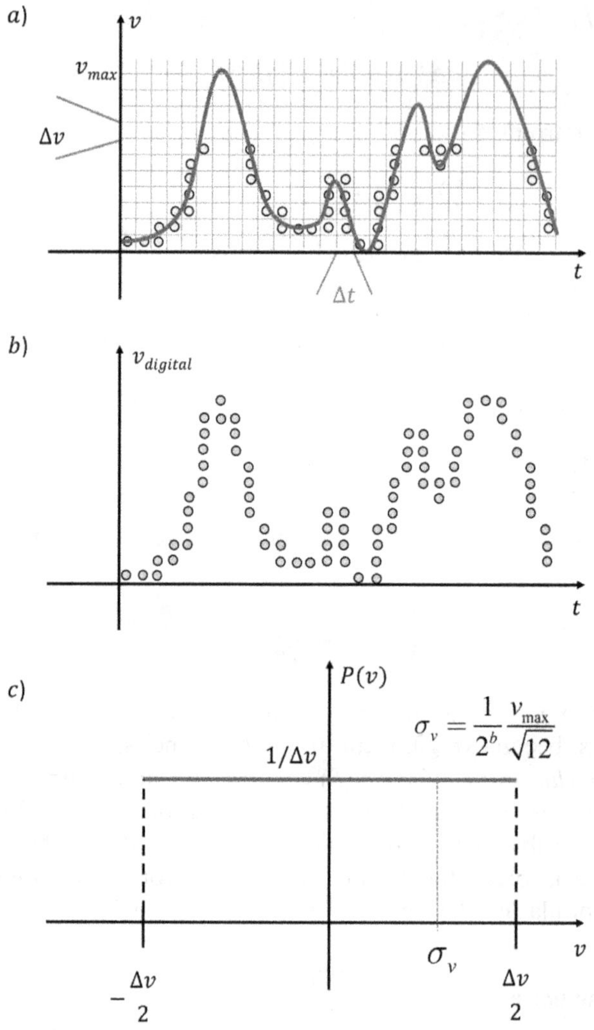

Figure 10.15. (a) Analog signal (solid blue line) sampled in time at intervals Δt and in ν at intervals $\Delta\nu$. (b) The resulting digital signal approximating the signal in (a). (c) Uniform distribution of ν values over the rounding interval $\Delta\nu$.

$$p(\nu) = \frac{1}{\Delta\nu}\prod\left(\frac{\nu}{\Delta\nu}\right) \tag{10.68a}$$

$$\int_{-\Delta\nu/2}^{\Delta\nu/2} p(\nu)\, d\nu = 1. \tag{10.68b}$$

In equation (10.68*a*), $\prod\left(\frac{\nu}{\Delta\nu}\right)$ is our rectangular function, introduced in Volume 1, section 4.4.

Because we chose the interval symmetric with respect to the origin, clearly, $\nu = 0$. As the digitizer returns only one value of ν within the interval $\Delta\nu$, the noise it introduces equals the standard deviation. σ_ν in this interval. The variance for this uniform distribution is (see also section 9.4)

$$\begin{aligned}\sigma_\nu^2 &= <\nu^2> \\ &= \int_{-\frac{\Delta\nu}{2}}^{\frac{\Delta\nu}{2}} \nu^2 p(\nu)\, d\nu \\ &= \int_{-\frac{\Delta\nu}{2}}^{\frac{\Delta\nu}{2}} \frac{\nu^2}{\Delta\nu}\, d\nu \\ &= \frac{\Delta\nu^2}{12}.\end{aligned} \tag{10.69}$$

We can conclude that the rounding noise is

$$\begin{aligned}\sigma_\nu &= \frac{\Delta\nu}{\sqrt{12}} \\ &= \frac{1}{\sqrt{12}}\frac{\nu_{\max}}{2^b}.\end{aligned} \tag{10.70}$$

Of course, the larger number of bits b that the A/D converter provides, the smaller the noise.

10.4 Problems

1. A noise signal is the sum of two independent sources, characterized by the power spectra

$$S_{1,2}(f) = a_{1,2}e^{-\frac{(f-f_{1,2})^2}{2\Delta f_{1,2}}}.$$

   Calculate the autocorrelation of the noise signal.
2. Show that the variance of noise in problem 1 is the sum of the individual variances.
3. Calculate the noise-equivalent bandwidth for the following power spectra:
   a) $S(f) = e^{-\frac{f^2}{2\Delta f}}$.
   b) $S(f) = \prod\left(\frac{f}{\Delta f}\right)$.
   c) $S(f) = e^{-\frac{|f|}{\Delta f}}$.
4. Calculate the *rms* voltage of a Johnson-dominated noise, if the resistor is $R = 1$ kΩ, at room temperature, at a bandwidth $\Delta f = 1$ MHz.
5. Under the shot noise regime, what is the average current that would generate a standard deviation of the photocurrent, σ_i, equal to the i_{rms} from problem 4, at the same temperature?

6. What is the SNR change factor for a shot noise dominated detector, when the same amount of optical power is delivered at a wavelength $\lambda_1 = 500$ nm versus $\lambda_2 = 600$ nm?
7. Under the shot noise regime, what is the maximum bandwidth for an SNR = 10, when the incident power is $P = 10^{-12}$ W, quantum efficiency is $\eta = 0.8$, and wavelength $\lambda = 1$ μm?
8. Calculate the photoconductive gain, G, that generates a generation–recombination *rms* current equal to that of shot noise, for an average photocurrent $i = 1$ nA and bandwidth $\Delta f = 1$ kHz.

References

[1] Wiener N 1930 Generalized harmonic analysis *Acta Math.* **5** 117–258
[2] Boyd R W 1983 *Radiometry and the Detection of Optical Radiation* (Wiley Series in Pure and Applied Optics) (New York: Wiley), vii p 254
[3] Johnson J B 1928 Thermal agitation of electricity in conductors *Phys. Rev.* **32** 97
[4] Nyquist H 1928 Thermal agitation of electric charge in conductors *Phys. Rev.* **32** 110
[5] Shottky W 1918 Über spontane Stromschwankungen in verschiedenen Elektrizitätsleitern *Ann. Phys.* **362** 541–67
[6] Lauritzen P O 1968 Noise due to generation and recombination of carriers in p-n junction transition regions *IEEE Trans. Electron Devices* **15** 770–6
[7] Dereniak E L and Boreman G D 1996 *Infrared Detectors and Systems* vol 306 (New York: Wiley)

IOP Publishing

Principles of Biophotonics, Volume 2

Light emission, detection, and statistics

Gabriel Popescu

Chapter 11

Figures of merit of optical detectors

When choosing an optical detector for a particular application, it is important to understand which characteristics make it a good detector. Here we review the most important figures of merit that allow us to compare various devices [1–4].

11.1 Quantum efficiency

Quantum efficiency of an optical detector is defined as the ratio between the number of photoelectrons, N_e, generated by a number of incident photons, N_q,

$$\eta = \frac{N_e}{N_q}. \tag{11.1}$$

For a given acquisition time, Δt, we can express η in terms of the incident photon flux, P_q (see section 3.3 for photon-based radiometric quantities) and a resulting photocurrent, i,

$$\begin{aligned} \eta &= \frac{P_q \Delta t}{i \Delta t / e}, \\ &= e\frac{P_q}{i}, \end{aligned} \tag{11.2}$$

where e is the electron charge. In the visible spectrum, Si-based detectors can yield $\eta > 90\%$, while in the near-IR spectrum, InGaAs detectors can have $\eta > 80\%$. Quantum efficiency is one of the most important figures of merit for an optical detector, as it defines the overall input–output response, as detailed in the next section.

doi:10.1088/978-0-7503-1644-6ch11

11.2 Responsivity

Responsivity describes the electrical *output* of a detector with respect to its *optical* input. The electrical output can either be voltage (in volts, V) or current (in amps, A). The responsivity in terms of photocurrent is defined as

$$R = \frac{i}{P} \tag{11.3a}$$

$$[R] = \frac{A}{W}, \tag{11.3b}$$

where P is the incident optical power (measured in W). For a constant responsivity, we can say that the detector's response is *linear*, $i \propto P$.

11.2.1 Spectral responsivity

We can express the responsivity as a function of optical *wavelength*, recalling the relationship between the photon flux and optical power, $P = h\nu P_q$ (section 3.3):

$$\begin{aligned} R(\upsilon) &= \frac{i}{h\upsilon P_q} \\ &= \frac{\eta e}{h\upsilon}. \end{aligned} \tag{11.4}$$

In arriving at equation (11.4), we used the expression for quantum efficiency in equation (11.2). Of course, an equivalent representation can be obtained in terms of the optical wavelength, $\lambda = c/\upsilon$,

$$R(\lambda) = \frac{\eta e}{hc}\lambda. \tag{11.5}$$

The responsivity increase with wavelength is because, for a given incident optical power, the lower energy photons (larger wavelengths) correspond to higher photon flux.

The quantum efficiency itself may depend on wavelength. The dependence $\eta(\lambda)$ is a characteristic of the material used (see figure 11.1(a)). Thus, the spectral responsivity incorporates this dependence as well, $R(\lambda)\alpha\eta(\lambda)\lambda$, which leads to nonlinear spectral responsivity curves (figure 11.1(b)). Figure 11.2 shows the responsivity for Si and InGaAs detectors.

11.2.2 Temporal responsivity

The *temporal* responsivity of a semiconductor detector, that is, how fast a detector responds to an optical signal (or how fast a signal can be modulated while still being detectable), depends on the kinetics of electron generation and recombination. In the first approximation, the rate of change of the number of photoelectrons, N, is

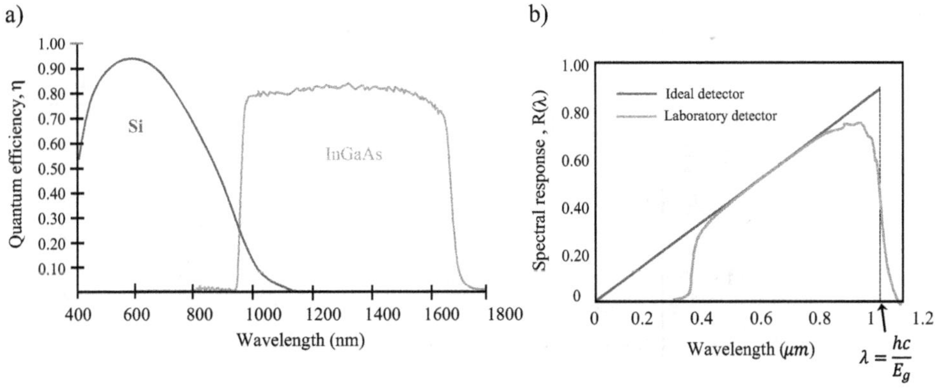

Figure 11.1. (a) Quantum efficiency for Si and InGaAs, as indicated. (b) Responsivity versus wavelength differs from the ideal linear curve.

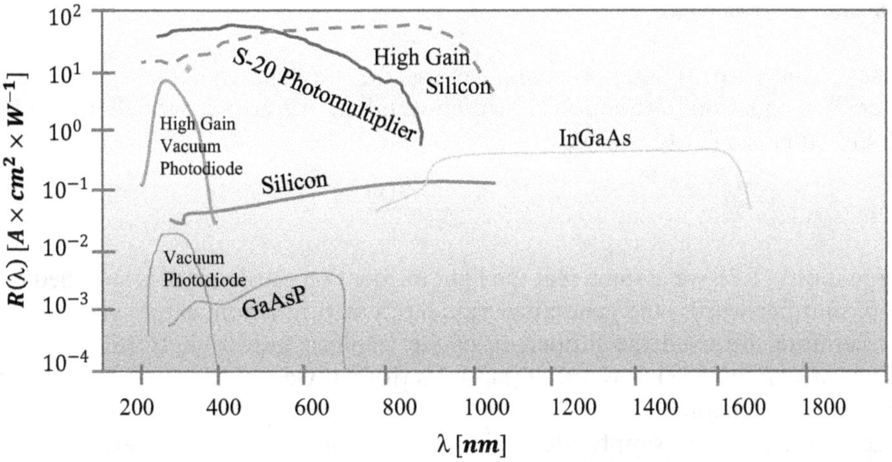

Figure 11.2. Spectral responsivity of various detectors.

$$\frac{dN(t)}{dt} = g(t) - \frac{N(t)}{\tau}; \tag{11.6}$$

where g is the electron–hole pair generation and τ is the photoelectron lifetime. Note that this equation is similar to the rate equation for the upper level in a two-level system, as encountered in laser kinetics (section 7.5, Example 1), in the absence of stimulated emission. The generation rate, g, is analogous to the pump and N/τ to the spontaneous decay. This analogy should not be surprising because carrier generation is indeed promoting an electron to a higher energy state, from the valence to the conductance band.

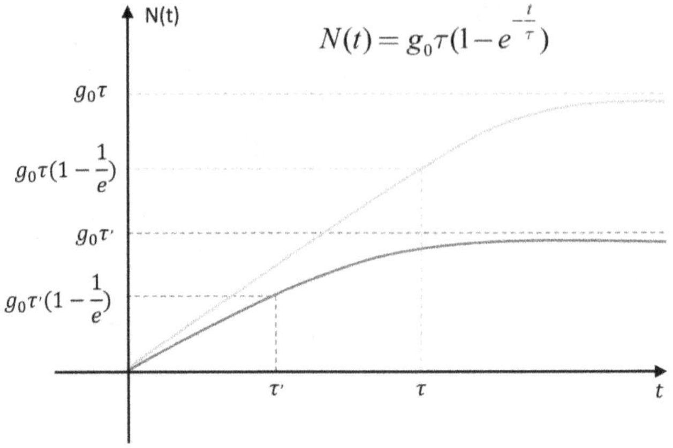

Figure 11.3. Yellow curve: number of photoelectrons as a function of time; the the light is switched on at $t = 0$ for a material of carrier lifetime τ. A shorter lifetime, $\tau' = \frac{\tau}{2}$, leads to faster response build-up, but saturates at a lower level, $g_0\tau'$ (green curve).

As in the case of laser kinetics, we use the Laplace transform to solve the differential equation in equation (11.6). Thus, taking the Laplace transform on both sides of equation (11.6), we have

$$sN(s) = \frac{g_0}{s} - \frac{N(s)}{\tau}. \tag{11.7}$$

In equation (11.7), we assume that the light incident on the detector is switched on at $t = 0$, in other words, the generation rate has a step function shape, $g(t) = \rho_0\Gamma(t)$. Furthermore, we used the properties of the Laplace transform from section 6.5, namely, $dN/dt = sN(s) - N(0)$, $\Gamma(t) \leftrightarrow 1/s$ (for a review of the Laplace transform, see Volume 1, chapter 11).

Equation (11.7) is simply algebraic in the Laplace frequency variable s. The solution is

$$\begin{aligned} N(s) &= \frac{g_0/s}{s + \frac{1}{\tau}} \\ &= g_0\tau\left[\frac{1}{s} - \frac{1}{s + \tau}\right]. \end{aligned} \tag{11.8}$$

We can easily take the Laplace transform inverse of equation (11.8) to bring N into the time domain, by invoking the shift theorem,

$$N(t) = g_0\tau(1 - e^{-\frac{t}{\tau}})\Gamma(t). \tag{11.9}$$

Equation (11.9) indicates that the detector requires a finite amount of time to reach its steady state, $N(t = \infty) = g_0\tau$ (figure 11.3). Interestingly, we see that a shorter lifetime leads to a faster ramp-up to the maximum value, but this happens as the maximum value (or steady state) is lower. Physically, this is the case because a shorter lifetime means a higher rate of recombination, which works against the carrier generation.

Next, let us investigate the situation where the incident light has both a switch-on and a switch-off moment. Specifically, in the previous example, we will consider that the light turns off after a duration T. Thus, the generation rate is now a rectangular function, which can be written as a difference of two shifted step functions.

$$g(t) = g_0[\Gamma(t) - \Gamma(t - T)], \tag{11.10}$$

which has a Laplace transform

$$g(s) = g_0\left(\frac{1}{s} - \frac{1}{s}e^{sT}\right).$$

The equation in $N(s)$ is obtained by taking the Laplace transform of equation (11.6), with the new $g(s)$, namely,

$$sN(s) = g_0\left(\frac{1}{s} - \frac{1}{s}e^{sT}\right) - \frac{N(s)}{\tau}. \tag{11.11}$$

Solving for $N(s)$, we obtain

$$\begin{aligned} N(s) &= \frac{g_0(1 - e^{sT})}{s\left(s + \frac{1}{\tau}\right)} \\ &= g_0\tau(1 - e^{sT})\left(\frac{1}{s} - \frac{1}{s + \frac{1}{\tau}}\right) \\ &= g_0\tau\left(\frac{1}{s} - \frac{1}{s + \frac{1}{\tau}}\right) - g_o\tau e^{sT}\left(\frac{1}{s} - \frac{1}{s + \frac{1}{\tau}}\right). \end{aligned} \tag{11.12}$$

We see on the RHS of equation (11.12) that the first term yields the previous solution, when g is a step function (or $T \to \infty$), and the second term differs by the exponential e^{sT}, which brings a shift in the time domain. Taking the inverse Laplace transform, we obtain $N(t)$ as

$$N(t) = g_0\tau(1 - e^{-\frac{t}{\tau}})\Gamma(t) - g_0\tau(1 - e^{-\frac{t-T}{\tau}})\Gamma(t - T). \tag{11.13}$$

Equation (11.13) is illustrated in figure (11.4). For $t \geqslant T$, the second term on the RHS kicks in and equation (11.13) becomes

$$\begin{aligned} N(t)|_{t\geqslant T} &= g_0\tau\left[e^{-\frac{t-T}{\tau}} - e^{-\frac{t}{\tau}}\right] \\ &= g_0\tau(1 - e^{-\frac{T}{\tau}})e^{-\frac{t-T}{\tau}}. \end{aligned} \tag{11.14}$$

Thus, once the light turns off, N decays exponentially from the level that it reached during carrier generation, which is $N(T) = g_0\tau(1 - e^{-\frac{T}{\tau}})$. We note that there is a

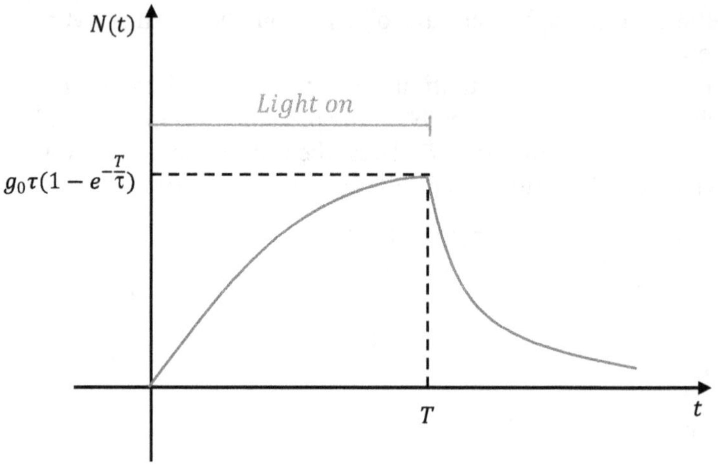

Figure 11.4. The number of electrons versus time, when the light is on for a duration T.

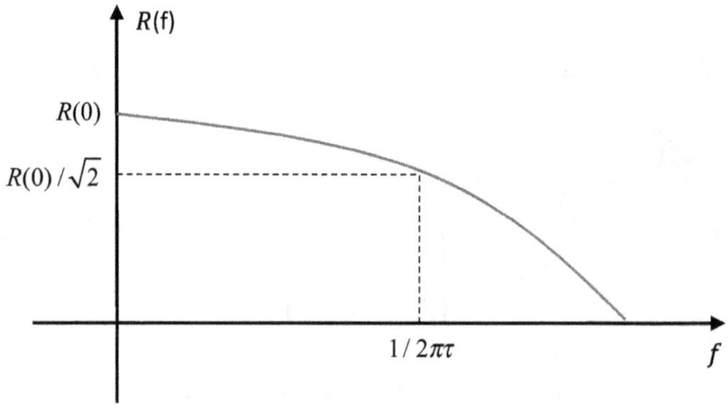

Figure 11.5. Typical detector temporal responsivity.

fundamental limit to how fast one can turn on and off the light, that is, to how short T can be: for $T \ll \tau$, the signal vanishes. Thus, the carrier lifetime ultimately limits the *temporal responsivity* of the detector.

For a simple exponential dependence of N, $N(t) = g\tau e^{-\frac{t}{\tau}}$, the Fourier transform is

$$N(f) = \frac{N(0)}{1 + i2\pi f\tau}. \tag{11.15}$$

Because the responsivity R is proportional to photocurrent, which itself is proportional to N, we can express temporal responsivity in terms of $|N(f)|$, namely

$$R(f) = \frac{R(0)}{\sqrt{1 + (2\pi f\tau)^2}}, \tag{11.16}$$

where $R(0)$ is the DC responsivity. $R(f)$ is sketched in figure 11.5. Figure 11.5 reiterates the fact that carrier lifetime determines the high frequency responsivity cut-off.

Finally, if we combine the *spectral* and *temporal* dependence of the *responsivity*, in other words, expressing $R(0)$ as a function of λ (equation (11.5))

$$R(0) = \frac{\eta(\lambda)e}{hc}\lambda, \tag{11.17}$$

we obtain the spectral and temporal responsivity $R(\lambda, f)$,

$$R(\lambda, f) = \frac{\eta(\lambda)e\lambda}{hc\sqrt{1 + (2\pi f\tau)^2}}. \tag{11.18}$$

Equation (11.18) establishes how sensitive a detector is for a certain wavelength and modulation frequency, f. Note that in experimental situations, the responsivity must be assessed relative to the noise level generated by the measurement.

11.3 Signal-to-noise ratio

SNR depends on both the detector and the input optical signal. The detector itself is characterized by a noise current, say, i_n, defined as *rms*. For a certain input optical power, P, the signal current, i_s, can be obtained using the responsivity, R_i,

$$i_s = R_i P. \tag{11.19}$$

Thus, the SNR is defined as

$$\begin{aligned} \mathrm{SNR} &= \frac{\mathrm{S}}{\mathrm{N}} \\ &= \frac{i_s}{i_n} \\ &= \frac{\mathrm{R}_i P}{i_n}. \end{aligned} \tag{11.20}$$

Clearly, as the signal current approaches i_n, it becomes difficult to detect (figure 11.6). Conventionally, SNR = 1 is considered to describe the lowest

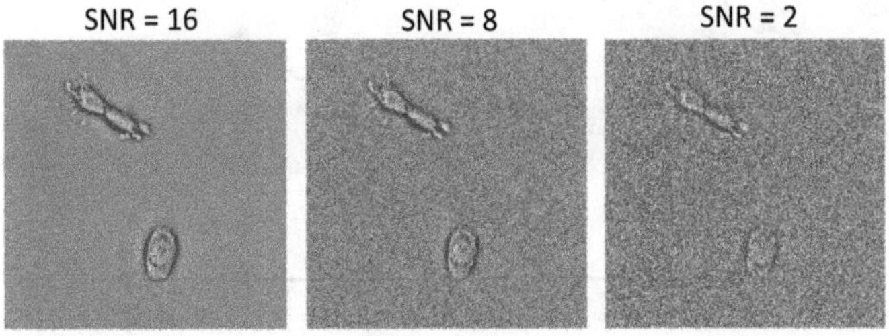

Figure 11.6. Illustration of an image detected at various values of SNR, as indicated.

detectable signal ($i_s = i_n$). Sometimes, SNR is defined as the signal at saturation versus noise at saturation. This way, SNR becomes a quantity only dependent on the detector and not on the input. *Saturation* is described next.

11.4 Saturation

Saturation describes the phenomenon by which the responsivity of the detector vanishes above a certain input value (figure 11.7). Thus, the detector cannot output currents above a certain saturation value, i_{sat}. Increasing the input power P will result in no change.

In an image sensor (CCD, CMOS), if all the pixels are saturated, we simply record a 'white' frame. The *saturation capacity,* or well depth, is typically expressed in electrons (e^-). The well depth of an image sensor represents the maximum number of electrons that a single pixel can store. Each pixel behaves like a bucket capable of holding electrons. The lower the number of electrons, the faster the camera saturates. The pixel size affects directly the saturation capacity.

11.5 Dynamic range

The dynamic range of a detector is the ratio between the saturation signal and the lowest signal that the camera can measure (i.e., the noise level),

$$DR = \frac{P_{\text{sat}}}{P_n}. \tag{11.21}$$

In equation (11.21), DR is the dynamic range, and P_{sat} and P_n are the incident optical power at saturation and noise level, respectively. If we assume a lower response across the entire dynamic range, that is, assuming a constant responsivity, R_i, the corresponding photocurrents are $i_{\text{sat}} = R_i P_{\text{sat}}$ and $i_n = R_i P_n$. Thus, DR can also be expressed as

$$DR = \frac{i_{\text{sat}}}{i_n}. \tag{11.22}$$

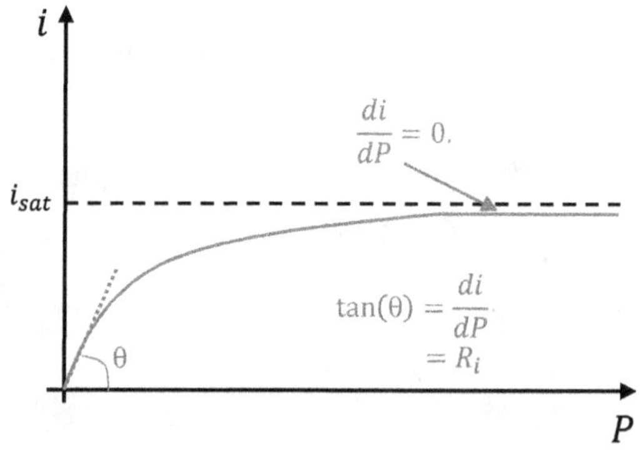

Figure 11.7. Photocurrent saturation in a photodetector.

Most commonly, DR is expressed in dB or bits,

$$DR_{\mathrm{dB}} = 10 \log\left[\frac{P_{\mathrm{sat}}}{P_n}\right] \tag{11.23a}$$

$$DR_{\mathrm{bit}} = \log_2\left[\frac{P_{\mathrm{sat}}}{P_n}\right]. \tag{11.23b}$$

Common cameras can have dynamic ranges of 8 bit ($DR = 256$), 10 bit ($DR = 4096$), and even 16 bit ($DR = 65\,536$). As shown in figure 11.8, a higher dynamic range allows for higher quality images, as dark and bright regions are revealed simultaneously.

11.6 Noise-equivalent power

Noise-equivalent power (NEP) describes the minimum optical power level that a given detector can record. Let us consider the current noise level, i_n, defined as the *rms* noise. If the responsivity of the detector is R_i, for an optical power incident, P, the signal current is

$$i_s = R_i P. \tag{11.24a}$$

Thus, we can define an SNR

$$\frac{S}{N} = \frac{R_i P}{i_n}. \tag{11.24b}$$

The NEP is defined as the power P for which SNR = 1, namely,

$$\mathrm{NEP} = \frac{i_n}{R_i}. \tag{11.25}$$

An analog expression is used for voltage-response detectors,

$$\mathrm{NEP} = \frac{v_n}{R_v}. \tag{11.26}$$

Since the responsivity is defined as the output signal current divided by the input power, $R_i = i_s/P$, NEP can also be expressed as

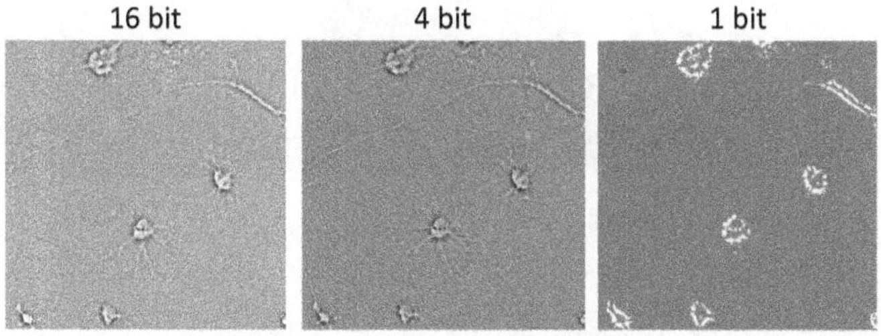

Figure 11.8. Effects of the dynamic range on the appearance of an image, as indicated.

$$\text{NEP} = \frac{P}{i_s/i_n} \qquad (11.27)$$
$$[\text{NEP}] = W.$$

Equation (11.27) indicates that NEP can be inferred for an arbitrary input power, by measuring the signal and noise currents.

Note that, as seen in section 11.2, the responsivity can have a spectral and frequency dependence, $R_i(\lambda, f)$, in which case NEP carries the same dependence,

$$\text{NEP}(\lambda, f) = \frac{i_n}{R_i(\lambda, f)}. \qquad (11.28)$$

In this case, $\text{NEP}(\lambda, f)$ is the power of a monochromatic light of wavelength λ that produces an SNR = 1, at frequency f.

Clearly, a low NEP describes a sensitive detector. We can describe *sensitivity* as 1/NEP. In general, NEP depends on other parameters, such as the detector area and noise equivalent bandwidth. Thus, a new figure of merit was introduced, to take these parameters into account, as described next.

11.7 Detectivity

Detectivity D^* is inversely proportional to NEP,

$$D^* = \frac{\sqrt{A_d \Delta f}}{\text{NEP}} \qquad (11.29a)$$

$$[D^*] = \text{m}\sqrt{\text{Hz}}/\text{W}. \qquad (11.29b)$$

D^* can be described as unit sensitivity, 1/NEP = 1, for a detector of unit area and unit noise equivalent bandwidth. Spectral and frequency dependent detectivity is defined via $\text{NEP}(\lambda, f)$,

$$D^*(\lambda, f) = \frac{\sqrt{Ad\Delta f}}{\text{NEP}(\lambda, f)}. \qquad (11.30)$$

11.8 Gain

Often, the current outputted by the detector is larger than the photocurrent directly produced by the incident light. The ratio between the output and the input currents is called *gain*, g_i. Of course, for voltage-generating detectors, we define a *voltage gain*, g_v,

$$g_i = \frac{i_{\text{out}}}{i_{\text{in}}} \qquad (11.31a)$$

$$g_v = \frac{v_{out}}{v_{in}}. \tag{11.31b}$$

The power is defined as $P = Ri^2 = \frac{v^2}{R}$, with R the impedance of the circuit. Thus, we can define a power gain, g_P, as

$$\begin{aligned} g_P &= \frac{P_{out}}{P_{in}} \\ &= \frac{R_{out}\, i_{out}^2}{R_{in}\, i_{in}^2} \\ &= \frac{v_{out}^2 / R_{out}}{v_{in}^2 / R_{in}}. \end{aligned} \tag{11.32}$$

Sometimes, the power gain is expressed in decibels (dBs), which is a logarithmic scale, and useful to describe large dynamic range signals,

$$g_P^{dB} = 10 \log g_P, \tag{11.33}$$

where log is a base-10 logarithm. A gain of a million (10^6) can be expressed more simply as 60 dB gain. Note that a factor of two gain represents 3 dB, while a gain of 1/2 gives −3 dB.

11.9 Dark current

The current that flows through the photodetector in the absence of light is called *dark current*. This current is due to the random generation of electron–hole pairs within the detector material. Dark current is one of the dominant sources of noise in imaging sensors, such as CCDs. Note that dark current is characterized by shot noise and, thus, cannot be completely removed by subtracting successive frames.

11.10 Spatial and temporal sampling: aliasing

A crucial property of a point-detector is how fast it can measure the signal, or how fast it can *sample* it. This temporal *sampling frequency* is $f = 1/\delta t$, in units of s^{-1}(Hz). For a 2D detector, in addition to the temporal sampling (frame rate), we also use *spatial sampling*, which is dictated by the pixel size. Spatial sampling frequency is $f_x = 1/\delta x$ (in m^{-1}). Figure 11.9 illustrates the sampling in the space and time of a signal.

It is instructive to see the effects of sampling in the frequency domain. Let us consider the input optical power, of spatial dependence, $P(x)$. If the pixel size is δx and the detector size is L, the sampled signal, P_s, is (figure 11.10)

$$P_s(x) = P(x)\mathrm{comb}\left(\frac{x}{\delta x}\right)\Pi\left(\frac{x}{L}\right), \tag{11.34}$$

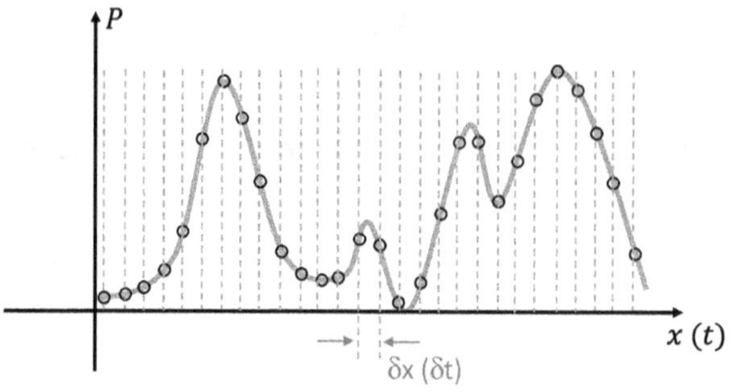

Figure 11.9. Spatial or temporal sampling by a detector: δx sampling in space (pixel size), δt sampling in time.

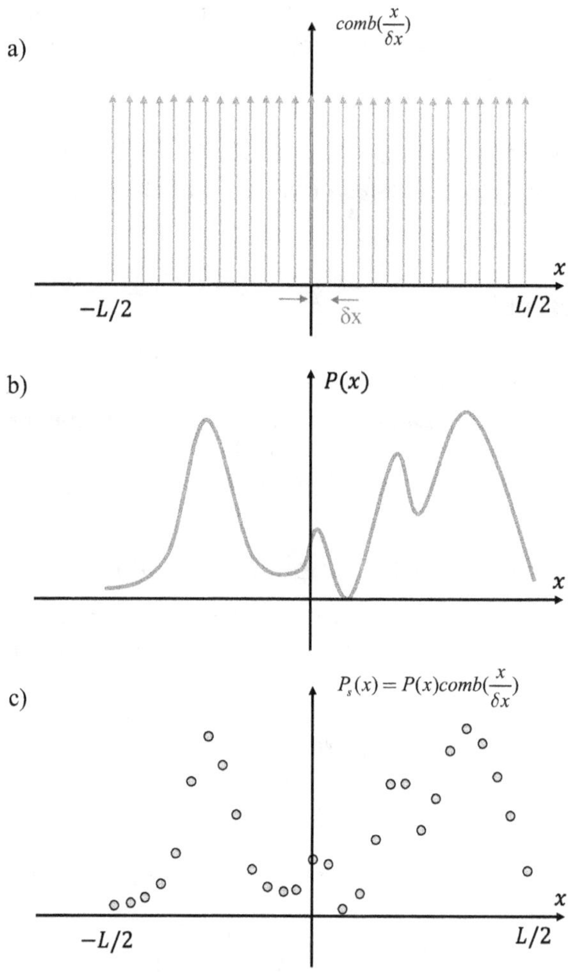

Figure 11.10. Spatial sampling of signal $P(x)$. (a) Array of delta-functions, of period δx. (b) Signal to be sampled, $P(x)$. (c) The resulting digitized signal.

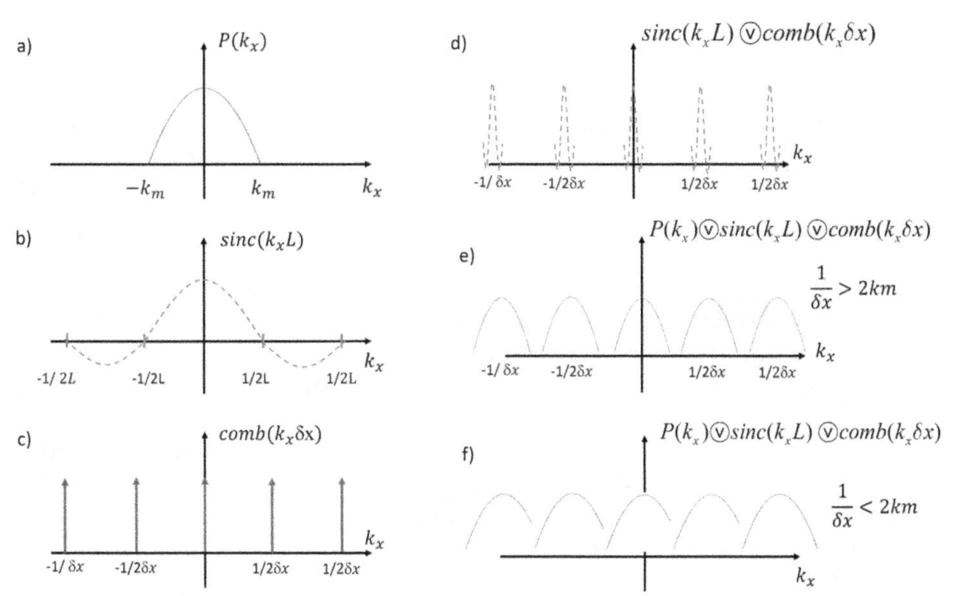

Figure 11.11. The effect of sampling in the frequency domain: (a) the original band-limited signal; (b) sinc(k_xL); c) comb($k_x\delta$x); (d) (sinc(k_xL)ⓥcomb($k_x\delta x$); (e) ($P(k_x)$ⓥsinc(k_xL)ⓥcomb($k_x\delta x$) with the Nyquist condition fulfilled; (f) ($P(k_x)$ⓥsinc(k_xL)ⓥcomb($k_x\delta x$) with the Nyquist condition *not* fulfilled (aliasing).

where comb$\left(\frac{x}{\delta x}\right)$ is the comb function of period δx and ⓥ stands for convolution, as described in Volume 1, chapter 4. It is instructive to investigate the result in the frequency domain. The Fourier transform of $P_s(x)$ is

$$P_s(k_x) = AP(k_x) ⓥ \text{sinc}(k_xL) ⓥ \text{comb}(k_x\,\delta x), \tag{11.35}$$

where A incorporates all the normalization constants.

Equation (11.35) indicates that the Fourier transform of the sampled signal $P_s(k_x)$ is very different from that of the original signal, $P(k_x)$,

$$P_s(k_x) = P(k_x) ⓥ f(k_x), \tag{11.36}$$

where $f(k_x) = A\,[\text{sinc}(k_xL) ⓥ \text{comb}(k_x\,\delta x)\,]$. Let us consider that the input signal is band limited, $P(k_x) = 0,\ for\ |k_x| \geqslant k_m = 2\pi f_m$. Thus, $P(k_x)$ and $P_s(k_x)$ are illustrated in figure 11.11.

We see that the sampling yields a periodic function in the frequency domain, with a period $k_x{}^0 = 2\pi f_x{}^0$. Figure 11.11(e) shows the situation when $k_x{}^0 > 2k_m$. In this case, the frequency content of the original signal repeats without overlap. This is nothing more than the Nyquist criterion, stating that the sampling frequency should be at least twice the highest frequency of the signal,

$$f_x^0 \geqslant 2f_m, \tag{11.37a}$$

which in terms of pixel size means

$$\delta x \leqslant 1/(2fm). \tag{11.37b}$$

However, if this criterion is not satisfied, there will be overlap in the periodic frequency pattern and the signal detection will yield errors (figure 11.11(f)). This under-sampling phenomenon is called *aliasing* and should be avoided at all costs.

11.11 Problems

1. X-ray radiation ($\lambda_1 = 0.1$ nm) is incident on a phosphorescent material, which creates green photons ($\lambda_2 = 550$ nm) with a power efficiency of conversion of $\alpha = 1\%$. A photodetector detects the green photons with a quantum efficiency $\eta_1 = 80\%$. What is the quantum efficiency, η_2, for detecting x-ray photons?
2. A 1 mW power of radiation at $\lambda = 1$ μm falls onto a detector of quantum efficiency $\eta = 0.7$. What is the photocurrent generated by the detector?
3. A photodetector has a responsivity R_1 at $\lambda_1 = 500$ nm, which is twice that at $\lambda_2 = 1.2$ μm. If the quantum efficiency at λ_1 is $\eta_1 = 0.9$, what is the quantum efficiency at λ_2?
4. A rectangular light pulse is incident on a semiconductor detector. The photodetector number kinetics $N(t)$ is characterized by equation (11.6). Compute $N(t)$ as a function of the electron–hole pair generation rate (which has the shape of the optical pulse), the photoelectron lifetime, and pulse width.
5. The optical power falling on a semiconductor detector is sinusoidally modulated, $P(t) = P_0(1 + \cos \Omega t)$, with Ω the frequency of modulation. Calculate $N(t)$ as a function of Ω, the electron–hole pair generation rate, g, and photoelectron life-time, τ.
6. The quantum efficiencies of a semiconductor detector at wavelengths λ_1 and λ_2 are η_1 and η_2, respectively. What should the ratio of the respective modulation frequencies, f_2/f_1, be, for which the responsivities are also equal?
7. A Johnson noise-dominated detector operates at room temperature and has a resistance of 10 $k\Omega$.
   a) Calculate the noise current at a modulation frequency of 100 kHz.
   b) Calculate the SNR for detecting 1 μW of red light ($\lambda = 633$ nm), if the quantum efficiency is $\eta = 0.9$.
   c) If the saturation optical power is $P_s = 15$ μW, calculate the dynamic range of the detector and express it in dB and bits.
8. The detector in problem 7 is cooled such that the thermal noise is negligible and becomes shot-noise limited.
   a) What is the shot noise-dominated noise current at 1 μW incident optical power ($\lambda = 633$ nm), $\eta = 0.9$?
   b) Calculate the SNR.
   c) Calculate the dynamic range and express it in dB and bits ($P_s = 15$ μW).

9. For the detector in problem 7, calculate the contribution of the generation–recombination noise. With the detector at room temperature, calculate:
   a) the total noise due to Johnson, shot, and generation–recombination noise;
   b) the new SNR;
   c) the new dynamic range, expressed in dB and bits.
10. What is the NEP for the detector in problem 7?
11. What is the detectivity of the detector in problem 9, if the detector has an area $A_d = 1\ \text{cm}^2$?
12. A CCD records an interferogram, that is, an irradiance distribution of the form $I(x, y) = I_0 + I_1(x, y) + 2\sqrt{I_0 I_1(x, y)}\cos[\alpha x + \phi(x, y)]$, where I_0 is a constant and I_1 and ϕ are band limited to k_m $\left(\sqrt{k_x^2 + k_y^2} \leqslant k_m\right)$.
   a) Calculate the minimum value of α that satisfies the Nyquist criterion for avoiding aliasing of the $\phi(x, y)$ signal.
   b) What is the largest pixel size that avoids aliasing of the signal $I(x, y)$?

References

[1] Ahmed S N 2007 *Physics and Engineering of Radiation Detection* 1st edn (Amsterdam: Academic), xxiv p 764

[2] Boyd R W 1983 *Radiometry and the Detection of Optical Radiation* (Wiley Series in Pure and Applied Optics) (New York: Wiley), vii p 254

[3] Dereniak E L and Boreman G D 1996 *Infrared Detectors and Systems* vol 306 (New York: Wiley)

[4] Kingston R H 1978 *Detection of Optical and Infrared Radiation* (Springer Series in Optical Sciences vol 10) (Berlin: Springer), viii p 140

Chapter 12

Semiconductor materials

12.1 Insulators and conductors

Photodetection is the process by which light is converted into a measurable quantity, typically an electrical signal. For example, the optical radiation can induce changes in the detector material such as resistance or inductance, which in turn produce changes in the measured current through or voltage across the detector (figure 12.1). During the detection process, a part of the incident optical power is converted into electrical power. Thus, photodetection is a *dissipative* process. In other words, if the light interacts with the material without exchange of energy, that is, *elastically*, detection is not possible.

Light–matter interaction is governed by the displacement of electronic charge that the optical field induces into the material. The incident light on a material applies an electric force on the electrons in that material,

$$\mathbf{F} = -e\mathbf{E}. \tag{12.1}$$

In equation (12.1), $-e$ is the charge of the electron and $\mathbf{E}$ is the electric field carried by the optical radiation. Figure 12.2 illustrates the interaction between the light and a *dielectric* (insulator) material. Insulators (e.g. glass) contain electrons tightly bound to the nuclei and the perturbation induced by the incident optical field happens without loss of energy. The electron behaves as a mass on a spring, with the electric force exerted by the nucleus acting as the restoring force. As a result, the incident light is re-radiated by the material with 100% of the original power recovered as transmitted or reflected light. This process explains why good dielectric materials are also transparent to electromagnetic radiation.

For *conductive* materials, the situation is rather different. In an ideal conductor, which is approximated well by a *metal*, electrons are free to move around. The binding to the nuclei is so weak ($F_N \cong 0$) that the electrons can be considered as a gas, moving unrestricted in the material. As a result, the electrons are accelerated by the incident optical field and undergo collisions with the atoms in the lattice

doi:10.1088/978-0-7503-1644-6ch12

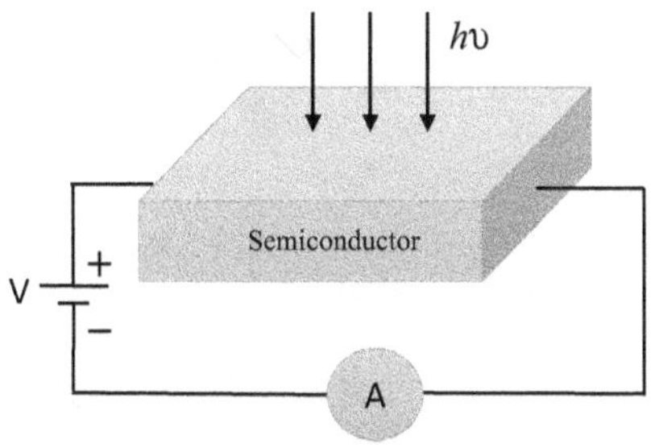

Figure 12.1. An example of photodetection mechanism whereby the incident light produces a change in the measured voltage across the detector or current flow.

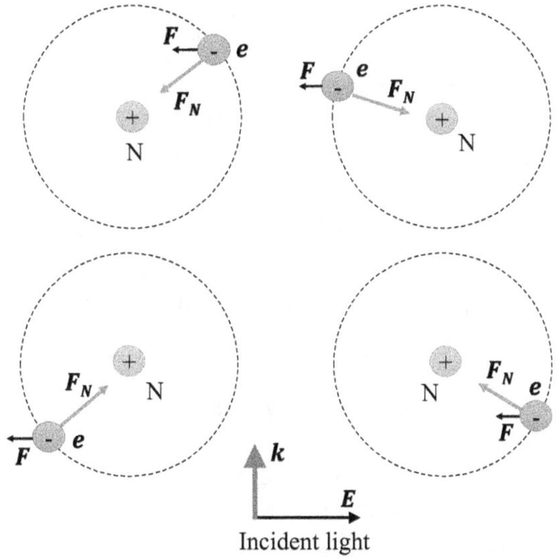

Figure 12.2. Interaction of the optical field, $\boldsymbol{E}$, wavevector $\boldsymbol{k}$, with an *insulator* (dielectric). Electrons (−) are bound strongly to the nuclei (+) by Force $\boldsymbol{F}_N$, such that the electric force, $\boldsymbol{F}$, applied by the incident field is unable to break them away. The light–matter interaction is *elastic* (no loss).

(figure 12.3). During the collision process, electrons transfer some of their energy to the vibrational modes of the lattice, which, eventually, converts into heat. Thus, the interaction of light with conductors is a *dissipative* process. This description explains why good conductors, such as metals, are generally *opaque* to electromagnetic radiation.

Semiconductors are materials with conductivity that falls between that of metals and insulators. Because their electrical properties can be tuned with respect to temperature, concentration of impurities, voltage bias, etc, semiconductors are commonly used in photodetectors. Next, we discuss the basic properties of semiconductor materials (for a classical reference on semiconductor devices, see [1]).

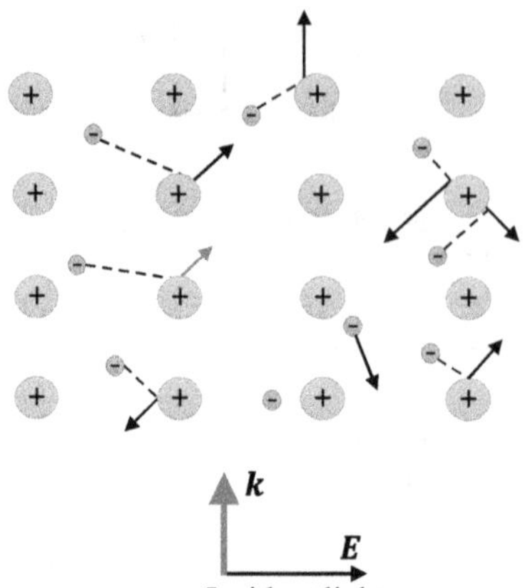

Figure 12.3. Interaction of light with a conductor. Electrons are free to move in the lattice and are accelerated by the electric field of incident light. Due to collisions with the ions in the lattice, electrons dissipate into heat some of the field energy.

12.2 Covalent bonds in semiconductor crystals

Semiconductors are solids in *crystalline form*, meaning that their atoms and molecules are arranged spatially in a regular and periodic manner (for a physical description of crystals, see [2]). A highly ordered three-dimensional distribution of atoms or molecules is known as a *crystal lattice*. In semiconductors, the atoms are held together by *covalent bonds*, in which pairs of electrons are shared between atoms (see figures 12.4 and 12.5).

Silicon (Si) and Germanium (Ge) are the most commonly used semiconductor materials. Both materials have four valence electrons, that is, four electrons on their outer, incomplete electronic shell. Silicon (Si, atomic number $Z = 14$) has a total of 14 electrons, 10 of which form complete shells ($1s^2 2s^2 2p^6 3s^2 3p^2$, in orbital notation). Germanium (Ge, $Z = 32$) has 32 electrons, 28 of which form complete shells ($1s^2 2s^2 2p^6 3s^2 3p^6 3d^{10} 4s^2 4p^2$, in orbital notation). The two electronic configurations are shown in figure 12.4.

Germanium's valence electrons exist on higher energy levels (farther from the nucleus) than those of silicon. As a result, the Ge electrons are more mobile, resulting in higher conductivity compared to Si.

Figure 12.5 illustrates how a Si atom can form covalent bonds with four other atoms, sharing a total of four pairs of electrons and, thus, creating a complete shell of electrons.

12.3 Energy band structure

The atomic model put forward by Niels Bohr in 1913 correctly explained the discrete spectral lines measured from the hydrogen atom. Bohr's atom consists of a condensed nucleus surrounded by revolving electrons, as illustrated in figures 12.4(a) and (b). However, this discrete occupancy of energy levels is not limited to isolated atoms. The covalently bound atoms in a semiconductor also

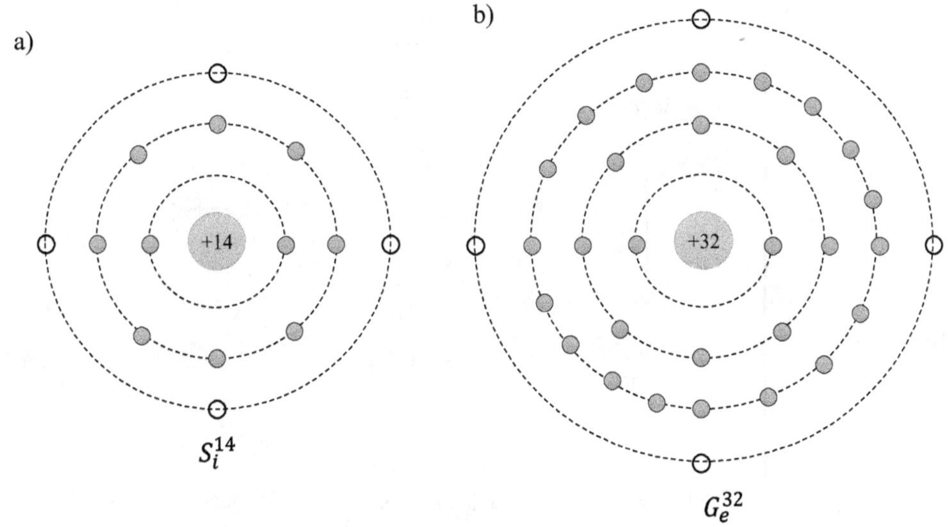

Figure 12.4. a) Electronic structure of Silicon (Si). Si has three electronic shells: $1s^2$ (two electrons), $2s^2$ $2p^6$ (eight electrons), $3s^2$ $3p^2$ (four electrons). b) Electronic structure of germanium (Ge). Ge has four electronic shells: $1s^2$ (2ē), $2s^2$ $2p^6$ (8ē), $3s^2$ $3p^6$ $3d^{10}$ (18ē), and $4s^2 4p^2$ (4ē). Both materials have four valence electrons, depicted by empty circles.

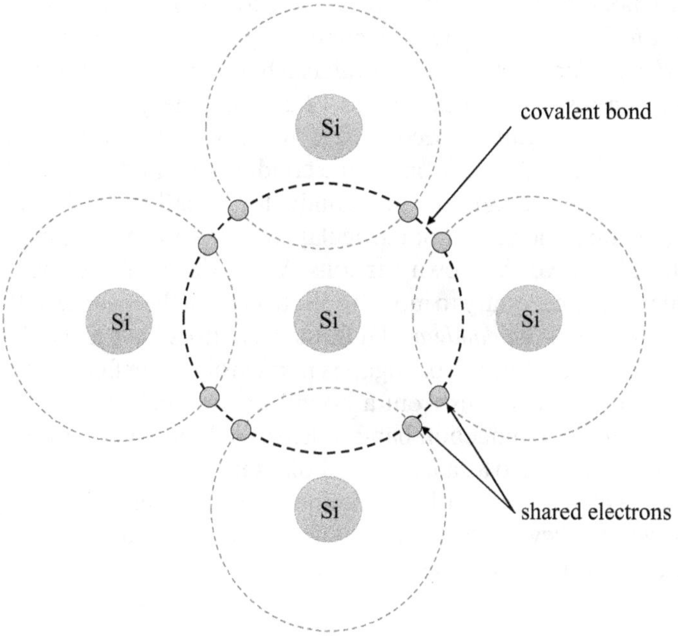

Figure 12.5. Illustration of the covalent bond in crystalline silicon. Pairs of electrons from the outermost shells are shared by an atom with four other atoms, creating a stable electronic shell (eight electrons).

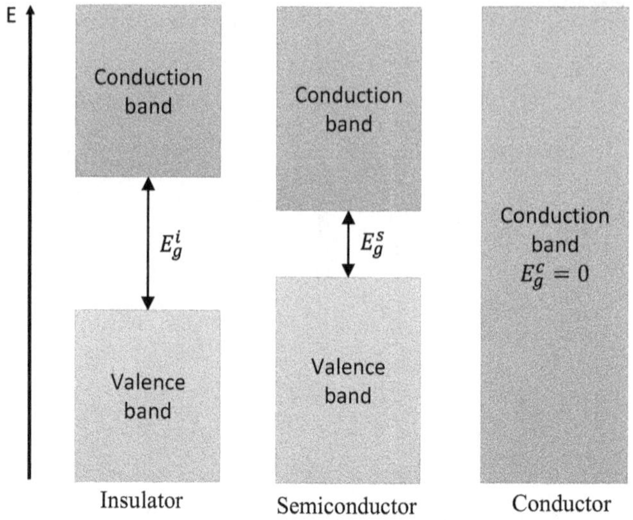

Figure 12.6. Energy band structure for insulators, semiconductors and conductors. The bandgap energy is much higher for insulators vs semiconductors, $E_g^i > E_g^s$, and non-existent for conductors, $E_g^c = 0$.

create discrete energy levels. However, unlike with isolated atoms, in a semiconductor crystal the energy levels are lumped into two bands: the *valence* and the *conduction* band (for a review of solid-state physics, see [3]).

The valence bond contains many closely packed energy levels. The two bands are separated by a *forbidden* energy region, containing no allowed energy levels, referred to as the *band gap*. The electrons in the valence band are tightly bound to the atoms, behaving as in an insulator (recall figure 12.2). They need to receive an energy at least equal to the bandgap to become free to move, that is, to transition to the conduction band. Thus, the electrons in the conduction band are loosely bound and virtually free to move around, like in a conductor (recall figure 12.3).

This energy band structure is not particular to only semiconductors, as conductors and dielectrics also have their own versions. What distinguishes the three types of materials is the energy necessary to move an electron from the valence to the conduction band, that is, the size of the *bandgap*. Thus, the insulators have a very large bandgap, while for an ideal conductor, the bandgap is non-existent (see figure 12.6).

When an electron transitions from a bound to free state, that is, when it moves from the valence to the conduction band, it leaves behind a net positive charge. This positive charge is referred to as a *hole*. The hole is not an actual particle containing a localized positive charge, but, rather, a fictitious particle defined by the absence of a negative charge. However, the concept of a hole is useful in describing charge transport in semiconductors.

12.4 Carrier distribution

Previously, when describing the black body radiation, we found that photons, being non-interacting, indistinguishable particles, or *bosons*, obey the Bose–Einstein

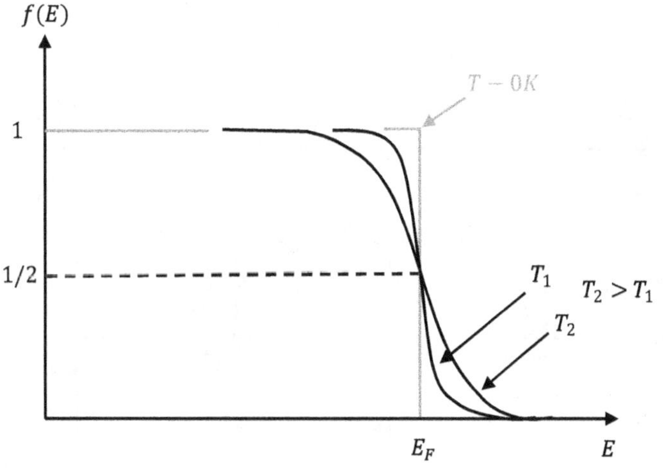

Figure 12.7. The Fermi–Dirac distribution.

statistics (see section 6.1). Thus, the probability of occupancy for a mode of energy $h\nu$ has the form (see Planck's formula in equation (6.2) and [4])

$$f(\nu) = \frac{1}{e^{\frac{h\nu}{k_B T}} - 1} \tag{12.2a}$$

where, as usual, k_B is the Boltzmann constant and T the absolute temperature. However, the energy distribution of electrons in a semiconductor is quantitatively different. Unlike photons, electrons obey Pauli's exclusion principle, which states that it is impossible for two electrons in an atom to have the same values of the four quantum numbers (principal quantum number, n, angular momentum quantum number, ℓ, magnetic quantum number, m_e, and m_s the spin quantum number). We discussed in section 5.1 how the electron spin defines the electronic state as a singlet, doublet or triplet.

Electrons belong to the family of *fermions*, particles that carry half-integer spin (note that photons have spin 1). As a result, the occupancy of the energy levels is governed by the Fermi–Dirac statistics (figure 12.7),

$$f(E) = \frac{1}{e^{\dfrac{(E - E_F)}{k_B T}} + 1}. \tag{12.2b}$$

In equation (12.2*b*), E_F denotes the *Fermi level.* To gain a physical understanding of the Fermi level, it is informative to study the case of $T \to 0$:

$$\lim_{T \to 0} e^{\frac{E - E_F}{k_B T}} = \left| \begin{array}{ll} 0, & if\ E < E_f \\ \infty, & if\ E > E_f \end{array} \right. . \tag{12.3}$$

Therefore, the occupancy probability at $T = 0\ K$ has the asymptotic values

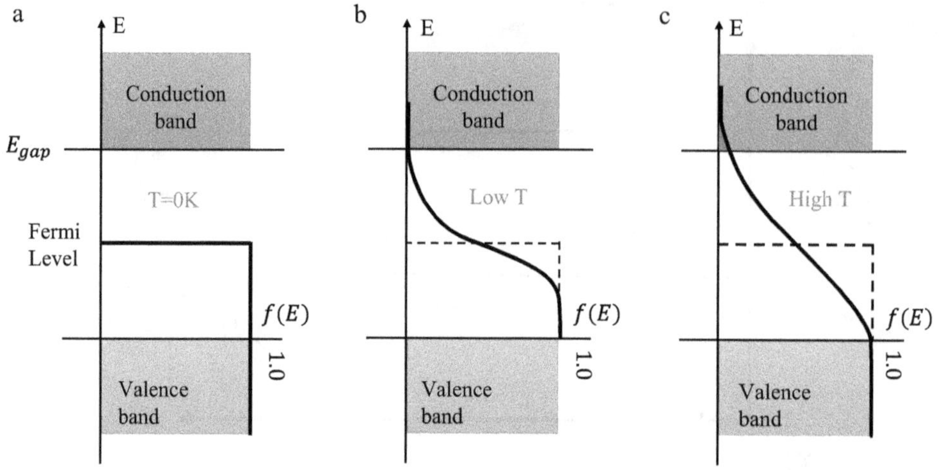

Figure 12.8. The Fermi distribution overlaid with the energy bands of an intrinsic semiconductor. (a) $T = 0$ K: no energy allowed above the Fermi level. (b) Low temperatures: energy levels above the Fermi level; only a few in the conduction band. c) High temperatures: some energy levels exist in the conduction band.

$$\lim_{T\to 0} f(E) = \begin{vmatrix} 1, & \text{if } E < E_f \\ 0, & \text{if } E > E_f \end{vmatrix}. \tag{12.4}$$

Equation (12.4) provides an insightful description of the Fermi level: it is the maximum energy occupied at absolute zero temperature. For intrinsic semiconductors, containing an equal number of positive and negative charge carriers, the Fermi level lies in the middle of the bandgap. This is the level at which the probability of occupancy is ½, $f(E_F) = 1/2$, meaning that the probability of occupancy at this energy level is ½. For silicon, $E_g = 1.12\, eV(1.79 \cdot 10^{-19}\text{ J})$, which indicates that the Fermi level lies $0.56\, eV$ above the valence band. For comparison, the thermal energy at room temperature is only $k_B T = 0.026\, eV$. This indicates that at $T = 300$ K, the probability of occupancy for energy levels $E > E_F$ is very low.

To gain a better understanding of the temperature effect, figure 12.8 shows $f(E)$ overlaid with the energy bands for a semiconductor, at zero, low, and high temperatures. The illustration in figure 12.8 provides information about the conductive properties of the semiconductor. Since there is a gap between the Fermi level and the conduction band, there will be absolutely no energy levels allowed in the conduction band at $T = 0$ K. It is important to note that even though $f(E)$ has finite values within the gap, there are still no electrons occupying those states, consistent with the definition of the gap. In order to understand this better, we calculate the number of particles per unit volume, $n(E)$, with energy within the interval $(E,\ E + dE)$. This quantity can be expressed as

$$n(E)dE = \rho(E)f(E)dE. \tag{12.5}$$

In equation (12.5), $\rho(E)$ is the density of states, meaning the number of energy states per unit volume within $(E,\ E + dE)$, while $f(E)$ is the Fermi distribution (equation (12.2b)). In order to calculate the density of states, we start by evaluating the number of modes in a cavity of volume V. Following the same reasoning as in section 6.1, the number of modes per wavenumber interval is (equation 6.15, chapter 6)

$$dN = \frac{V}{2\pi^2} k^2 \, dk. \tag{12.6}$$

To change the variable from wavenumber to energy, we use the relationships from quantum mechanics,

$$E = \frac{p^2}{2m} \tag{12.7a}$$

$$\mathbf{p} = \hbar \mathbf{k} \tag{12.7b}$$

where $\mathbf{p}$ is the momentum of the electron, $p = [\mathbf{p}]$, m its mass, and $\hbar = h/2\pi$ the reduced Planck's constant. Combining equations (12.7a and b), we obtain the following $k - E$ relationship

$$k = \left(\frac{2mE}{\hbar^2}\right)^{1/2} \tag{12.8a}$$

$$dk = \frac{1}{\hbar}\sqrt{\frac{2m}{E}} \, dE. \tag{12.8b}$$

If we now plug equations (12.8a and b) into equation (12.6), we obtain

$$\begin{aligned} dN &= \frac{V}{2\pi^2} \frac{2mE}{\hbar^2} \frac{1}{\hbar} \sqrt{\frac{2m}{E}} \, dE \\ &= \frac{4\pi (2m)^{3/2}}{h^3} V \sqrt{E} \, dE. \end{aligned} \tag{12.9}$$

Thus, we can now express the density of states present in equation (12.5) as

$$\begin{aligned} \rho(E) &= \frac{1}{V} \frac{dN}{dE} \\ &= \frac{4\pi (2m)^{3/2}}{h^3} \sqrt{E}. \end{aligned} \tag{12.10}$$

Using equations (12.5) and (12.10), we can now express the number of particles in the conduction band per unit volume and energy within the interval $(E,\ E + dE)$, namely, $n(E) = \rho(E) f(E)$. Semiconductor and conductor materials are of course, characterized by the same Fermi distribution, $f(E)$. However, the density of states for semiconductors is at the top of the gap, in other words, ρ is shifted by the bandgap energy, E_g. For a conductor, the density of states starts at the

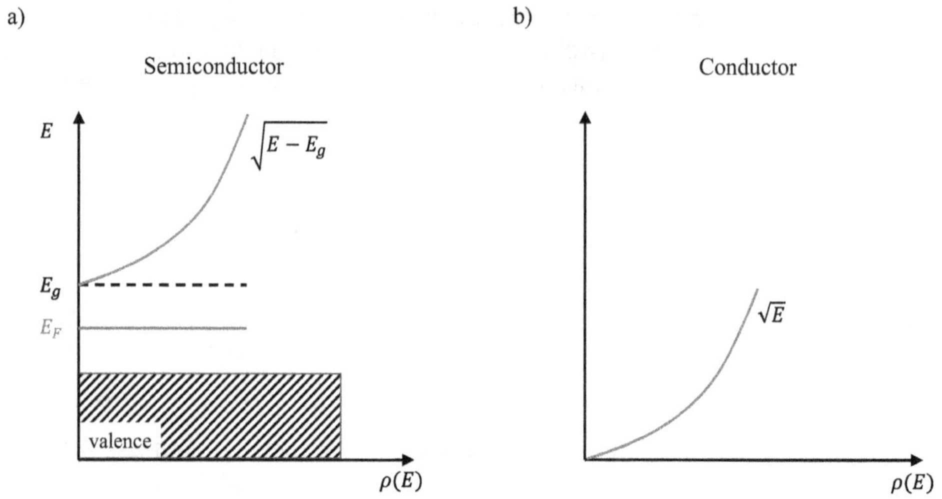

Figure 12.9. Density of states starts at the top of the bandgap for a semiconductor (a), and at the bottom of the valence band for a conductor (b).

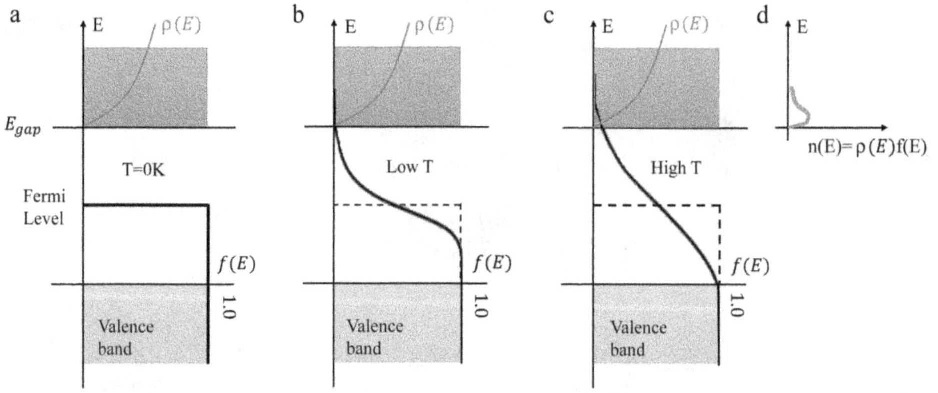

Figure 12.10. (a–c) Multiplying the density of states (figure 12.8) with the Fermi distribution yields the carrier distribution (electron population), $n(E)$. (d) Carrier distribution for the high temperature situation shown in (c).

bottom of valence band (see figure 12.9). Thus, we have the following expressions for the electron population in the conduction band

$$n(E) = \frac{4\pi(2m)^{3/2}}{h^3}\frac{1}{e^{\frac{E-E_F}{k_BT}}+1}\sqrt{E-E_f}\text{, for semiconductors} \tag{12.11a}$$

$$n(E) = \frac{4\pi(2m)^{3/2}}{h^3}\frac{1}{e^{\frac{E-E_F}{k_BT}}+1}\sqrt{E}\text{, for conductors.} \tag{12.11b}$$

Finally, we are ready to update figure 12.7 with the electron population curves, as shown in figure 12.10(d).

In order to find out the total number of electrons in the conduction band, n_c, we integrate $n(E)$ from the bottom of the conduction band to infinity,

$$n_c = \int_{E_g}^{\infty} n(E)dE. \tag{12.12}$$

For silicon and germanium at energies in the conduction band, the following inequality holds, $E - E_F \geqslant k_BT$. This approximation is well justified if we note that, at room temperature ($T = 300k$)

$$k_BT = 0.026\, eV \tag{12.13a}$$

$$E_g^{Si} = 1.1\, eV \tag{12.13b}$$

$$E_g^{Ge} = 0.67\, eV. \tag{12.13c}$$

As a result of this approximation, $e^{\frac{E-E_F}{k_BT}} + 1 \simeq e^{\frac{E-E_F}{k_BT}}$, and

$$\begin{aligned} \frac{1}{e^{\frac{E-E_F}{k_BT}} + 1} &\simeq e^{-\frac{(E-E_F)}{k_BT}} \\ &= e^{-\frac{E_g}{2k_BT}}. \end{aligned} \tag{12.14}$$

In equation (12.14), we used the fact that, for an intrinsic semiconductor, E_F lies in the middle of the bandgap and $E - E_F = E_g/2$. With this approximation, equation (12.11a) simplifies to

$$n(E) = \frac{4\pi(2m)^{3/2}}{h^3} e^{-\frac{E_g}{2k_BT}} \sqrt{E - E_f}. \tag{12.15}$$

Integrating over the energy, equation (12.12) yields for the electron concentration in the conduction band,

$$n_c = AT^{3/2}e^{-E_g/2k_BT} \tag{12.16a}$$

$$\begin{aligned} A &= \frac{2(2\pi mk_B)^{3/2}}{h^3} \\ &= 4.83 \times 10^{21}\, \frac{\text{electrons}}{\text{m}^3\text{k}^{3/2}}. \end{aligned} \tag{12.16b}$$

Equations (12.16a–b) show that, at room temperature, Si has $n_c^{Si} = 1.4 \times 10^{16}$ electrons/m^3. On the other hand, Ge has $n_c^{Ge} = 5.9 \times 10^{19}$ electrons/m^3, a significantly larger number.

12.5 Doping

So far, we described the energy band structures and carrier distributions for *ideal* semiconductors. These materials are assumed to be free of impurities. This class of semiconductors are called *intrinsic.* In such materials, the electrons and holes are

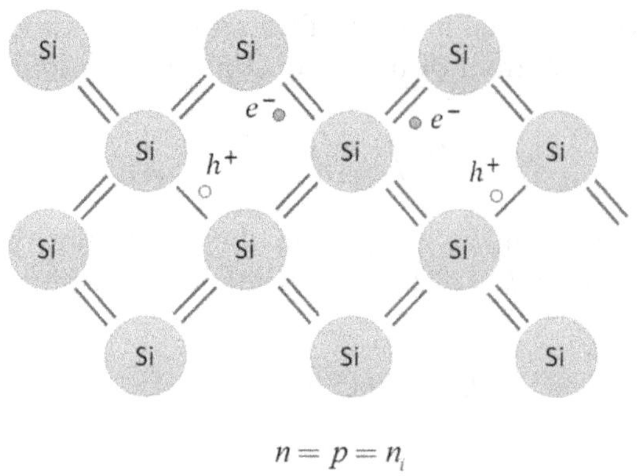

Figure 12.11. Electron–hole pairs in a Si crystal: e^-, electron, h^+, hole. The concentration of electrons is equal to that of holes in an intrinsic semiconductor.

created in pairs. Thus, the concentration of electrons (n) in the conduction band equals that of the holes (p) in the valence band,

$$n = p = n_i \tag{12.17}$$

where n_i is the carrier concentration for an intrinsic semiconductor. Figure 12.11 illustrates the electron–hole pair (EHP) generation in an intrinsic semiconductor.

Recombination is the reverse process of EHP *generation*. The rate of generation, g_i, $[g_i]$ = EHP/m^{3}s and of recombination, r_i, $[r_i]$ = EHP/m^{3}s, must be equal, at any temperature. The rate of recombination is proportional to the equilibrium concentrations of both electrons (n_0) and holes (p_0),

$$\begin{aligned} r_i &= cn_0p_0 \\ &= cn_i^2 \\ &= g_i, \end{aligned} \tag{12.18}$$

where c is a proportionality constant that depends on the specifics of the recombination process.

The electrical properties of semiconductors can be modified to accomplish particular tasks if impurities are added to the intrinsic material. This process is referred to as *doping* and is the most common method for tuning the conductivity of semiconductors. Through doping, a material receives either an excess of electrons, thus becoming an *n-type* material, or excess of holes, for a *p-type* material. As a result, the equilibrium concentrations, n_0, p_0, no longer equal the intrinsic carrier concentration, n_i.

N-type semiconductors are typically obtained by adding impurities from column V of the periodic table (figure 12.12a). These elements have five electrons on their outer shell, of which only four can form covalence bonds with the crystal (recall figure 12.5). Therefore, the fifth electron is weakly bound and can participate in conduction. These elements from column V are called *donor impurities*, because they contribute free electrons to the material.

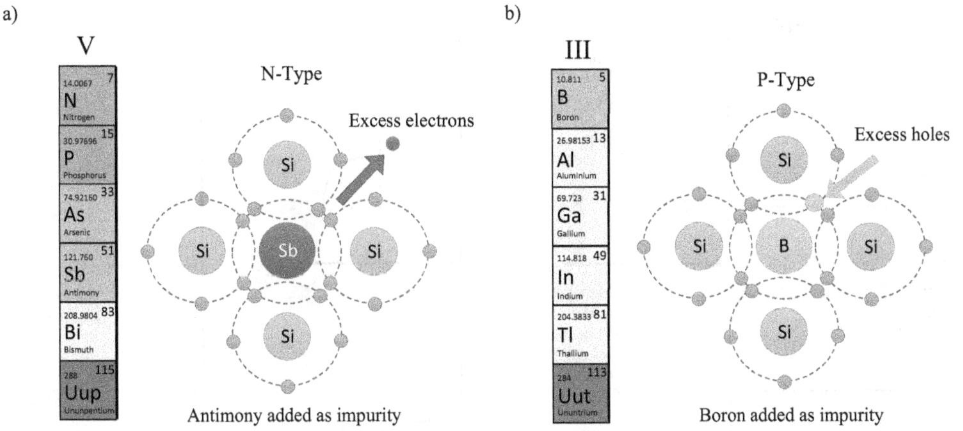

Figure 12.12. (a) N-type semiconductors are obtained by adding impurities from column V of the periodic table. These elements contribute free electrons. (b) P-type semiconductors are doped with elements from column III, which create holes.

P-type semiconductors are obtained by doping with elements from column III of the periodic table (figure 12.12b). In this case, the three electrons on the outer shell of these atoms are not sufficient to create the covalent bonds with the crystal. Thus, one bond remains incomplete as it misses one electron, or possesses an extra hole. We anticipate that in p-type semiconductors, there is excess of hole concentration (p) in the *valence* band.

Doping brings modifications to the band diagram, density of states, Fermi–Dirac distributions, and the carrier concentrations (see figure 12.13). For an n-type material (figure 12.13b), due to the excess of electrons, the Fermi level is shifted up, closer to the conduction band. As a result, the electron concentration is increased in the conduction band, at the expense of the hole concentration in the valence band. In p-type semiconductors, the Fermi level is shifted down, such that the hole concentration is increased in the valence band. The electron concentration in the conduction band is reduced.

12.6 Electron–hole pair generation by absorption of light

The fundamental process involved in photodetection is the generation of EHPs through absorption of radiation. These EHPs are often called *excess carriers*, indicating that they add to the existing carrier concentration at thermal equilibrium. Since the generated EHPs are out of equilibrium with their environment, they must eventually recombine.

As illustrated in figure 12.14, a photon with energy above the bandgap of the material, $h\nu > E_g$, can be absorbed, creating an electron in the conduction band and one hole in the valence band. As the valence band contains available electrons and the conduction band has numerous available energy states, the absorption process has high probability. The electron excited to the conduction band may have an energy higher that most electrons and will eventually lose this excess energy via

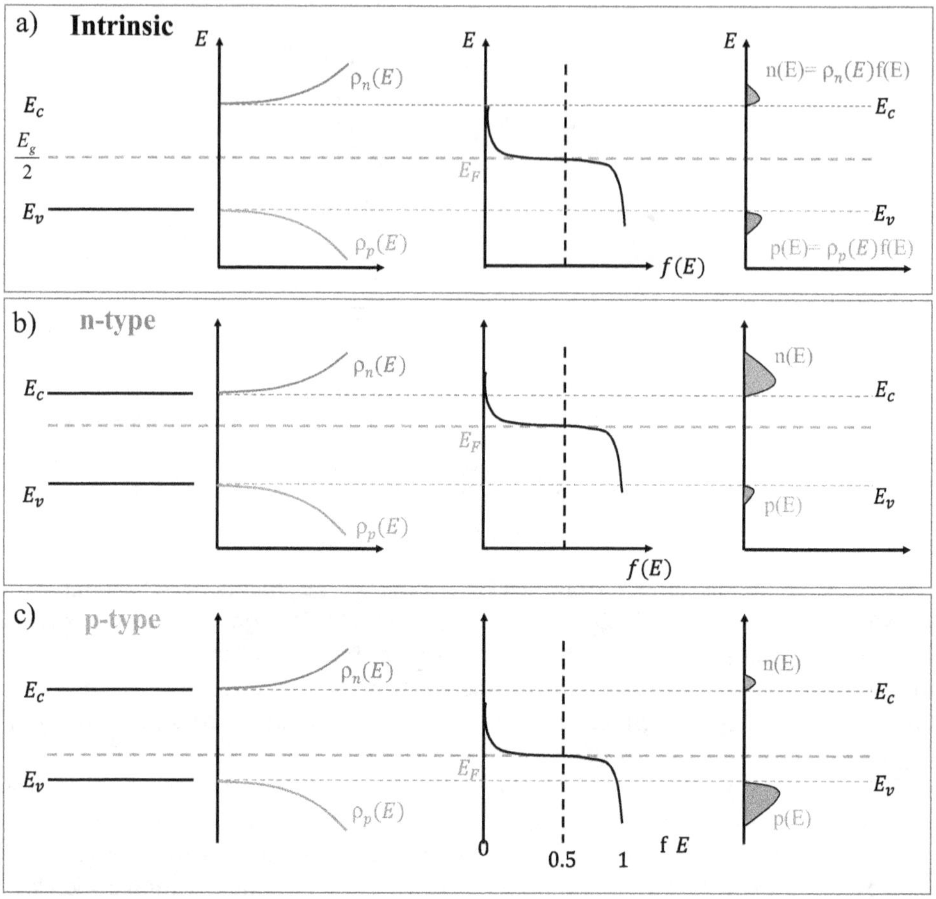

Figure 12.13. Band diagram, density of states (ρ), Fermi–Dirac distribution (f), and the carrier concentrations (n, p): (a) intrinsic, (b) n-type, and (c) p-type semiconductors at thermal equilibrium.

scattering with the lattice. This dissipative process excites vibrations in the lattice, which eventually converts into heat. As a result, the electron will lower its velocity and reach an energy level close to E_c. Finally, the electron can *recombine* with a hole in the valence band. This recombination process can be accompanied by the emission of a photon, a process called *photoluminescence*. Luminescence can also occur as a result of material bombardment with high-energy electrons (*cathodoluminescence*) or running a current through the material (*electroluminescence*). Photons with energies below the bandgap cannot excite electrons to the conduction band and, thus, are not absorbed. The semiconductor is *transparent* to photons of energies $h\nu < E_g$ and cannot act as detector at these wavelengths.

Let us consider a plane wave of *photon irradiance* I_q^0(in s^{-1}/m^2) incident on to a semiconductor of thickness L (figure 12.15). The change in I_q due to the absorption in a slice of thickness dz is proportional with the photon irradiance at the slice, $I_q(z)$, and the thickness of the slice,

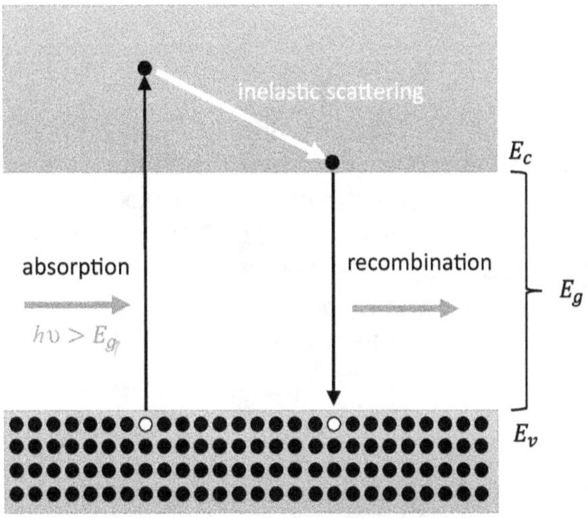

Figure 12.14. Optical absorption of a photon with energy higher than the bandgap creates an EHP. The electron excited in the conduction band loses energy via inelastic scattering with the lattice and occupies a lower energy level in the conductance band. Finally, the electron recombines with a hole in the valence band and can produce photoluminescence.

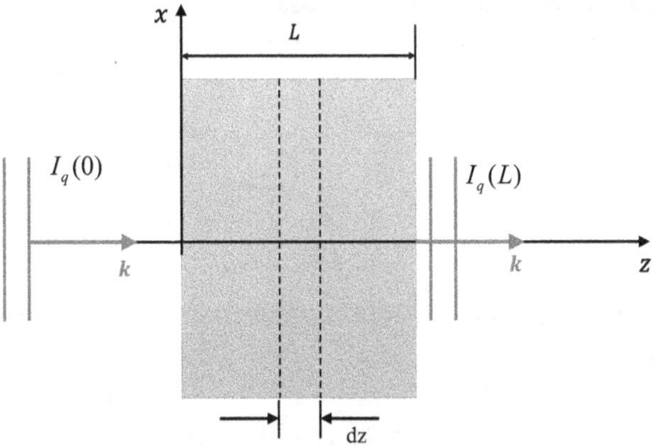

Figure 12.15. Lambert–Beer law for absorption of light in a material of thickness *L*.

$$dI_q(z) = -\alpha I(z)dz. \tag{12.19}$$

In equation (12.19), the proportionality constant, α, is called the *absorption coefficient* and the negative sign denotes a decrease in I_q with propagation, as expected. To obtain the photon irradiance at the exit surface of the material, $z = L$, we integrate equation (12.19),

$$\int_0^L \frac{dI_q(z)}{I_q(z)} = -\alpha z \Bigg|_0^L, \tag{12.20}$$

which yields

$$I_q(L) = I_q(0)e^{-\alpha L}. \tag{12.21}$$

Equation (12.21) represents the Lambert–Beer law of absorption.

Note that the absorption coefficient (units of m^{-1}) has a wavelength dependence that depends on the material. As discussed, we expect very low absorption for photon energies below the bandgap. Figure 12.16(a) illustrates the α versus $h\nu$ dependence and figure 12.16(b) shows the bandgap energies for various materials and the corresponding wavelengths. There is a useful relationship that connects the photon energy in units of eV and its wavelength,

$$E(\text{eV}) = \frac{1.24}{\lambda(\mu\text{m})}. \tag{12.22}$$

As shown in figure 12.16(b), many materials have bandgap energies below the visible range. For example, Si has a bandgap $E_g^{\text{S}_\text{i}} = 1.17\text{eV}$, which corresponds to

$$\begin{aligned} \lambda_g^{\text{S}_\text{i}} &= \frac{1.24}{1.17}\mu\text{m} \\ &= 1.06\ \mu\text{m}. \end{aligned} \tag{12.23}$$

Thus, Si makes an excellent photodetector material for visible light $\lambda \in (0.4, 0.75)\ \mu\text{m}$, but becomes essentially transparent for $\lambda > 1\ \mu\text{m}$. Germanium, on the other hand, has a smaller bandgap, $\lambda_g^{\text{Ge}} = 0.67\ \text{eV}$; this can detect longer wavelengths, up to

$$\begin{aligned} \lambda_g^{\text{G}_\text{e}} &= \frac{1.24}{0.67}\ \mu\text{m} \\ &= 1.83\ \mu\text{m}. \end{aligned} \tag{12.24}$$

For detecting even deeper into IR, one can use *InSb*, of bandgap $E_g^{InSb} = 0.17\ \text{eV}$, which gives

$$\begin{aligned} \lambda_g^{InSb} &= \frac{1.24}{0.74}\ \mu\text{m} \\ &= 7.3\ \mu\text{m}. \end{aligned} \tag{12.25}$$

Similarly, to detect UV light, we have to use materials of larger bandgap. For example, Z_nS has a bandgap, $E_g^{Z_nS} = 3.54\ \text{eV}$, which yields

$$\begin{aligned} \lambda_g^{Z_nS} &= \frac{1.24}{3.54}\ \mu\text{m} \\ &= 0.35\ \mu\text{m}. \end{aligned} \tag{12.26}$$

Figure 12.16(c) shows the absorption coefficient of various materials as a function of wavelength. Various stoichiometries of InGaAs are very popular for building

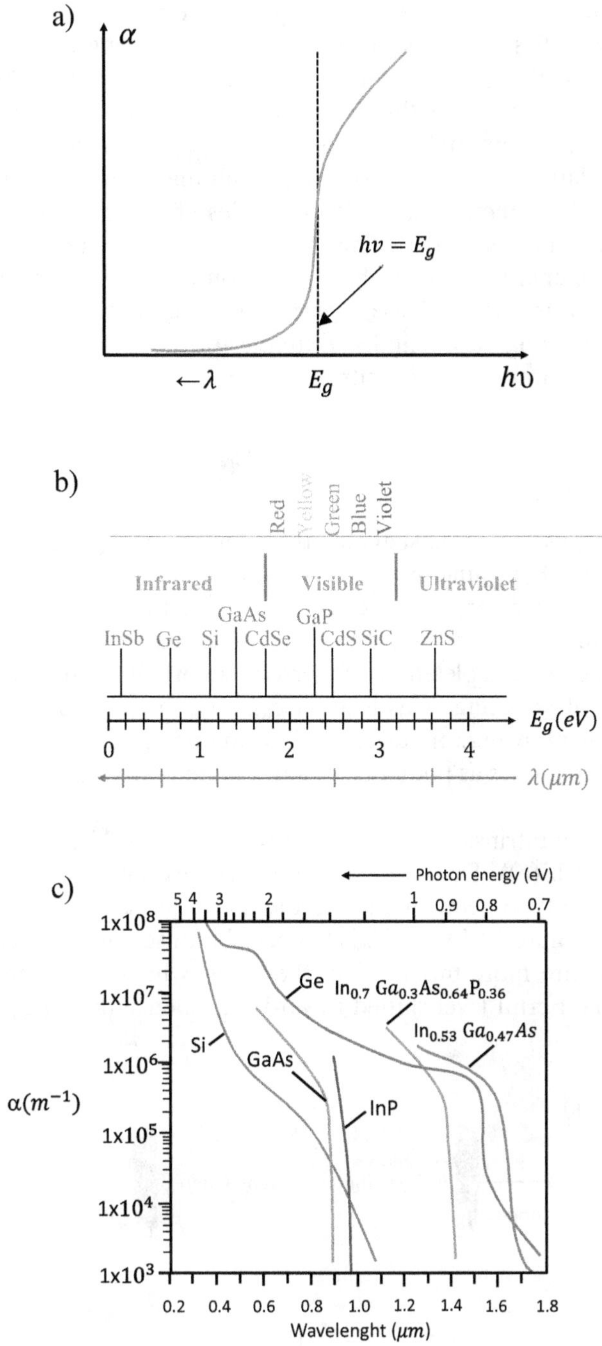

Figure 12.16. (a) Absorption coefficient increases abruptly for energies above the bandgap. (b) Energy bandgap and respective wavelengths for various materials. (c) Wavelength dependence of the absorption coefficient for various materials.

photodetectors in the near infrared. The largest wavelength detectable, corresponding to the bandgap, is sometimes called the *cut-off* wavelength.

Cooling is a valuable procedure to boost the performance of detectors, particularly at long cut-off wavelengths. By cooling, the number of carriers generated thermally goes down. Thermal excitation happens by the incident photon exciting vibrations on the lattice (excites photons) which ultimately increases the temperature in the material. This temperature rise increases the probability of an electron occupying the conduction band. This thermal excitation results in noise (dark current). Since the energy levels in the conduction band are distributed as $e^{-E_g/2k_BT}$ (see section 12.4), this thermal noise is particularly significant for E_g materials, that is, for detection of long wavelengths. Extreme forms of cooling, called *cryogenic* cooling, by, for example, liquid nitrogen ($T = 77$ K) brings the thermal noise essentially to zero.

12.7 P–N junction

A p–n junction is, as the name suggests, the interface between a p-type and n-type semiconductor [5]. P–n junctions are broadly used in semiconductor devices, including diodes, transistor, solar cells, LEDs, and integrated circuits. Although both p-doped and n-doped materials are somewhat conducting, the junction between the two can be depleted of carriers via recombination and, thus, rendered nonconductive, unless voltage bias is applied across the junction. The junction can operate as a *diode*, allowing for current to flow in one direction and not in the other (see section 12.8 on diodes as photodetectors). Figure 12.17 illustrates a junction and its circuit symbol.

Recall that for an intrinsic semiconductor the Fermi level lies in the middle of the bandgap (figure 12.13(a)). The Fermi level is the energy level filled with a probability of 50%. For an n-type semiconductor, the Fermi level is raised closer to the conduction band (figure 12.13(b)). This happens because the majority of the carriers are electrons and are more mobile than the holes, which are trapped. The p-type materials have the Fermi level shifted toward the valence band (figure 12.13(c)).

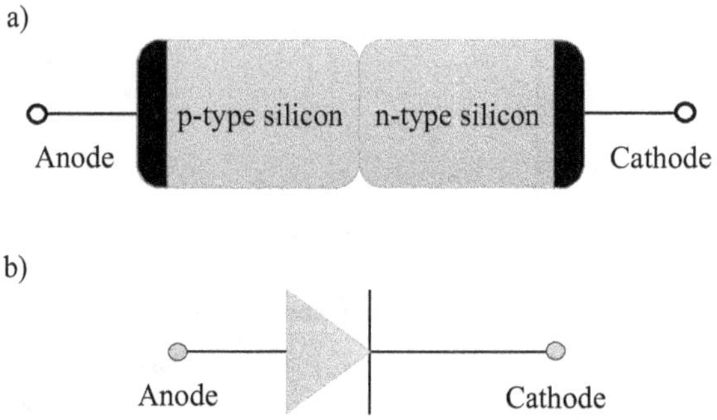

Figure 12.17. (a) A p–n junction formed by connecting a p-type and an n-type Si material. (b) Diode symbol: the base of the triangle corresponds to the p (*anode*) side.

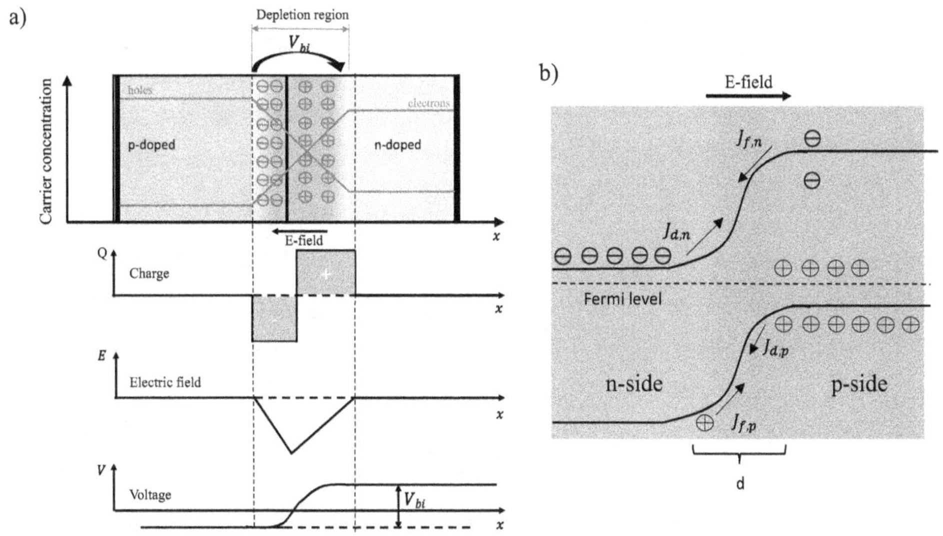

Figure 12.18. (a) A p–n junction in thermal equilibrium with zero-bias voltage applied. Under the junction, plots for the charge density, the electric field, and the voltage are shown. (b) The band diagram for the p–n junction, indicating that the Fermi levels overlap across the two regions. J indicates particle flux, the underscript d indicates diffusion flux, while f stands for "field" (directed) flux due to the electric field **E**. As usual, e and h stand for electrons and holes, respectively. D is the width of the depletion region.

Forward-and reverse-bias operation correspond to placing the positive voltage at the anode and cathode, respectively, and allow the junction to operate as a diode. First, we study the p–n junction in the absence of bias.

12.7.1 Zero bias

Figure 12.18(a) illustrates a p–n junction at thermal equilibrium and zero bias. The free electrons in the n-type material are attracted by the p-side, where they recombine with the holes and neutralize. However, the donor dopants in the n-type material are fixed and remain positive. Conversely, the acceptor dopants in the p-type material remain negatively charged. Thus, at equilibrium, there is a potential difference, known as the *built-in potential*, V_{bi} (see figure 12.18(a)).

The electric field, **E**, generated by charge build-up at the interface tends to oppose the diffusion of both the electrons and holes. Fick's law states that the diffusion flux, say for electrons, J_n, is proportional to the particle concentration gradient,

$$\mathbf{J}_n = -D_n \nabla n(\mathbf{r}) \tag{12.27}$$

where D_n is the diffusion coefficient for electrons and n their concentration. For holes, Fick's law has the analog form, namely,

$$J_{d,p} = -eD_p \frac{dp(x)}{dx}, \tag{12.28}$$

where $J_{d,p}$ is the hole diffusive current (see figure 12.18(b)), flowing from the p- to the n-side, e is the elementary charge, p is the concentration of holes, assumed to only vary in 1D, along x, and D_p is the diffusion coefficient of holes.

The current generated by the electric field E represents a *drift* current, $J_{f,p}$ (figure 12.18(b)) defined as

$$\mathbf{J}_{f,p} = e\mu_p p(x)\mathbf{E}, \tag{12.29}$$

where μ_p is the mobility of holes. At equilibrium, the two currents are equal, that is,

$$e\mu_p p(x)E = eD_p \frac{dp(x)}{dx}. \tag{12.30}$$

Integrating equation (12.30), we obtain

$$\mu_p \int_0^d E\, dx = D_p \int_{p_-}^{p_+} \frac{dp(x)}{p(x)}, \tag{12.31}$$

where d is the width of the depletion region, p_- is the hole concentration in the n-side and p_+ the hole concentration in the p-side. Performing the integrals in equation (12.31), we obtain for the built-in voltage

$$V_{bi} = \frac{D_p}{\mu_p} \ln\left(\frac{p_+}{p_-}\right). \tag{12.32}$$

In equation (12.32), $V_{bi} = \int_0^d E\, dx$ is the built-in voltage (see figure 12.18(a)). In 1905, Einstein derived an expression for the diffusion coefficient of charged particles at thermal equilibrium, as

$$D = \frac{k_B T \mu}{q}, \tag{12.33}$$

where, as usual, k_B is Boltzmann's coefficient, T is the absolute temperature, and q is the charge of the particle. Thus, the built-in voltage can be expressed by combining equations (12.32) and (12.33), as

$$V_{bi} = \frac{k_B T}{e} \ln\left(\frac{p_+}{p_-}\right). \tag{12.34}$$

Equation (12.34) gives an expression for the voltage at equilibrium, when the diffusion and drift currents cancel out for both the holes and electrons.

12.7.2 Forward bias

In *forward* bias, the positive electrode is attached to the p-type material and the negative electrode to the n-type material (see figure 12.19(b)). In this configuration, the positive voltage repels holes from the p-side and the negative voltage repels electrons from the n-side. The overall effect is that the depletion region shrinks. If

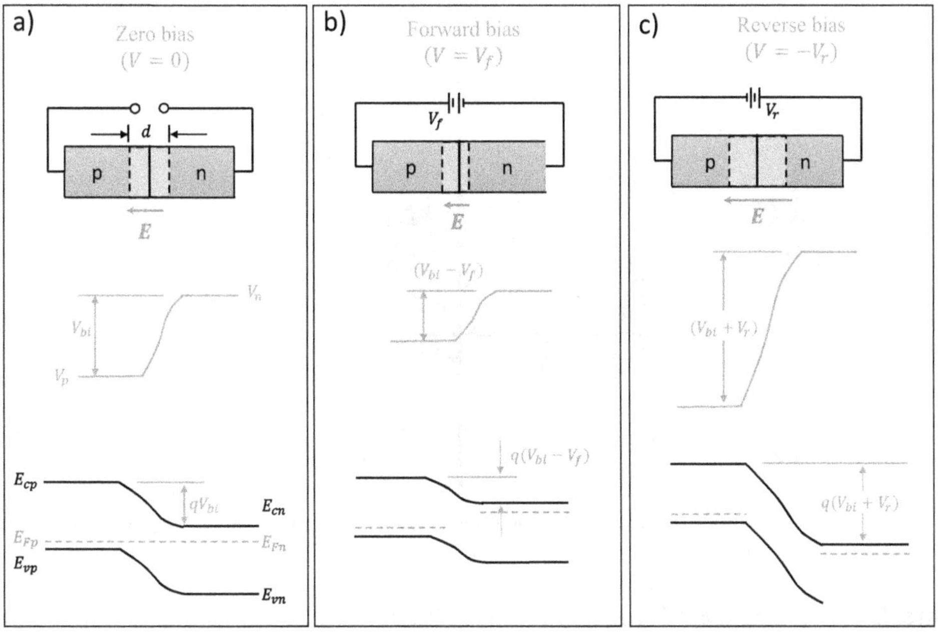

Figure 12.19. p–n junction, voltage, and energy levels for zero-bias (a), forward bias (b), and reverse bias (c).

the forward-bias voltage is progressively increased, the depletion region may become so thin that the built in electric field is not large enough to generate a drift current to cancel the diffusive current. As a result, the overall resistance of the junction decreases. In this *diffusion-dominant* regime, the electrons in the p-region and holes in the n-region eventually recombine, on average, after a distance called *diffusion length,* typically of the order of microns. Note that, although these *minority carriers* penetrate only a short distance in the material, the majority carrier current insures constant charge flow. Thus, in the p-region, the holes travel in the opposite direction to the electrons and, since their charge is also different in sign, contribute to the same current. The converse situation happens in the n-region. Thus, increasing the forward-bias voltage results in a sharp increase in current (see figure 12.20 for the current versus voltage diagram).

12.7.3 Reverse bias

The reverse bias of the p–n junction is achieved by connecting the negative electrode to the p-region and vice versa (figure 12.19(c)). Because the negative electrode tends to pull the holes away from the junction, the depletion region tends to increase in width (a similar effect happens in the n-region). Overall, this phenomenon leads to an increased resistance in the junction, which acts roughly as an insulator. Increasing the reserve bias voltage, that is, applying more negative voltage, yields a high built-in electric field, but without significant increase in the current. At some point, once the electric field intensity reaches a critical value, the p–n junction depleted zone

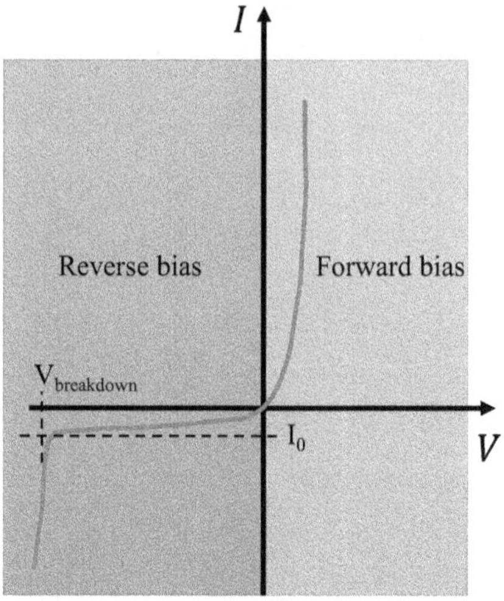

Figure 12.20. Characteristic *I–V* curve for a p–n junction: I_0, reverse saturation current, $V_{\text{breakdown}}$, breakdown (avalanche or Zener) voltage.

'breaks down' and the current begins to flow (figure 12.20). This phenomenon is used in *avalanche* or *Zener* diodes.

12.8 Problems

1. Using the Fermi distribution (equation (12.2b)) show that the probability for an electron to occupy the Fermi level is ½. Find the energy levels for which the probability is 0.1 and 0.9.
2. Calculate the intrinsic carrier concentration for Si at $T = 100$ K, $T = 200$ K, and $T = 300$ K.
3. Calculate the intrinsic carrier concentration for Ge at $T = 100$ K, $T = 200$ K, and $T = 300$ K.
4. Calculate the relative change in intrinsic charge concentration in Si versus Ge, $n_{\text{Si}}/n_{\text{Ge}}$, for a temperature change from $T = 300$ K to $T = 260$ K.

References

[1] Sze S M and Ng K K 2006 *Physics of Semiconductor Devices* (Hoboken, NJ: Wiley)

[2] Nye J F 1984 *Physical Properties of Crystals: Their Representation by Tensors and Matrices* (Oxford: Oxford University Press)

[3] Kittel C 2005 *Introduction to Solid State Physics* (New York: Wiley)

[4] Streetman B G and Banerjee S 2006 *Solid State Electronic Devices* 6th edn (Prentice Hall Series in Solid State Physical Electronics) (Upper Saddle River, NJ: Pearson/Prentice Hall), xviii p 581

[5] Chuang S L 1995 *Physics of Optoelectronic Devices* (Wiley Series in Pure and Applied Optics) (New York: Wiley), xv p 717

IOP Publishing

Principles of Biophotonics, Volume 2
Light emission, detection, and statistics
Gabriel Popescu

Chapter 13

Photon detectors

13.1 The p–n junction photodiode

13.1.1 Principle of operation

The p–n junction (see figure 13.1, and [1–4]) is capable of detecting optical radiation, as the incident photons create additional current flow. The incident photon must have sufficient energy to cause an electron to transition from the valence to the conductance band.

A photodiode is designed to operate in reverse bias. The optically active region of the photodiode is limited to the diffusion lengths of the electrons and holes (figure 13.1),

$$L_n = \sqrt{\frac{k_B T \mu_n \tau_n}{e}} \tag{13.1a}$$

$$L_p = \sqrt{\frac{k_B T \mu_p \tau_p}{e}}, \tag{13.1b}$$

where $\mu_{n,p}$ are their respective carrier mobilities, $\tau_{n,p}$ are the lifetimes, and e is the charge of the electron.

Let us consider the diode illuminated by a constant irradiance, which generates g electron–hole pairs (EHPs) per unit volume, per unit time. The EHPs on the n-side, for example, create minority holes. If those carriers are generated within a diffusion length (equation (13.1)) away from the boundary ($x = 0$, in figure 13.1(b)), a diffusion current will be registered at the diode terminals. The total current crossing the junction is the sum of the diffuse and drift currents.

At equilibrium, $V = 0$, the net current is zero, as the drift current cancels the diffusion component,

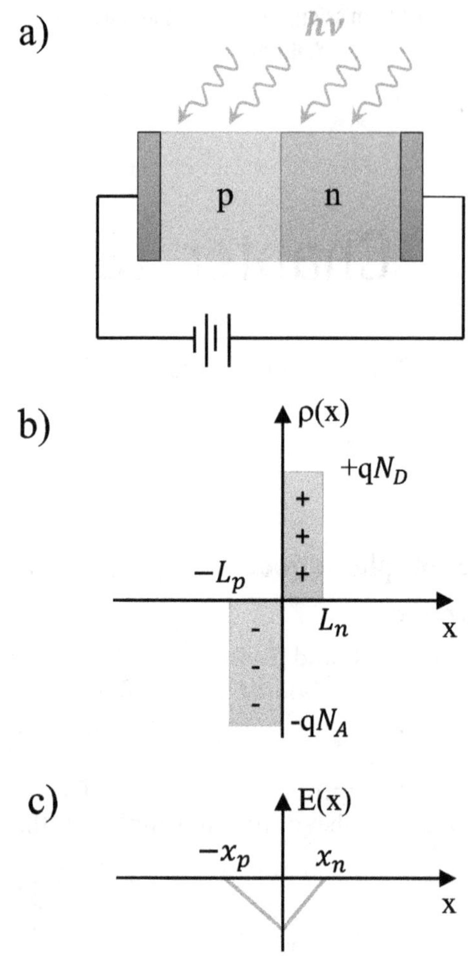

Figure 13.1. (a) A p–n junction diode under illumination. (b) The charge distribution $\rho(x)$ in the depletion region. (c) E is the electric field, $E(x)$ across the depletion region.

$$\begin{aligned} I &= I_{\text{diff}} - I_{\text{drift}} \\ &= 0. \end{aligned} \tag{13.2}$$

With a bias voltage V, either positive or negative, the probability of the carrier to cross the junction goes up by a factor of $\exp(eV/k_{\text{B}}T)$. Thus, the I–V curve is described by,

$$I = I_0(e^{eV/K_{\text{B}}T} - 1), \tag{13.3}$$

where, as in figure 12.20, I_0 is the saturation current. Equation (13.3) is illustrated in figure 13.2. For large values of the voltage, $\exp(eV/K_{\text{B}}T) \gg 1$, and the dependence approaches an exponential. At large negative V-values (reverse-bias), when $e^{-e|V|/k_{\text{B}}T} \ll 1$, the current approaches the *reverse saturation current* $I = -I_{\text{o}}$.

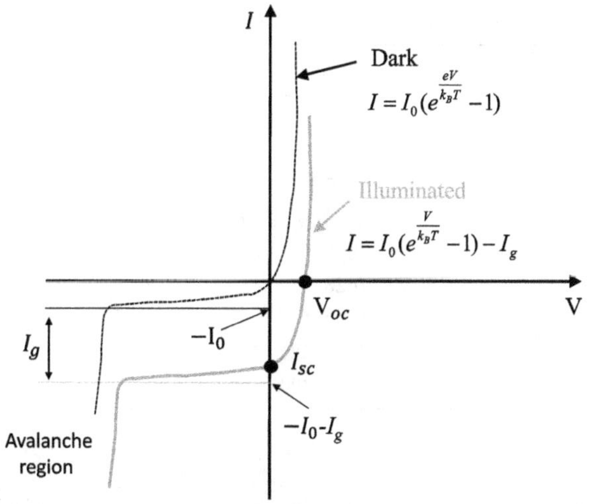

Figure 13.2. Photodiode characteristic curve with and without illumination: V_{oc} is the open circuit voltage, I_{sc} the short circuit current, I_g the photo-generated current, and $-I_0$ the reverse saturation current. The steep increase in the negative current is due to breakdown, which defines the avalanche region.

When light is incident on the p–n diode, there is a photocurrent generated, of the form (figure 13.2)

$$I_g = -eAG_0(L_p + L_n), \tag{13.4}$$

where A is the transverse area of the junction, G_0 is the number of photo carriers generated per unit volume, per unit time, and $L_{p,n}$ are the diffusion lengths defined in equations (13.1a–13.1b). As a result, the total current across the junction is

$$I(V) = I_0(e^{eV/k_BT} - 1) - eAG_0(L_p + L_n). \tag{13.5}$$

Equation (13.5) is illustrated in figure 13.2. Note that the I–V curve of the photodiode is affected by the incident optical power, P, because $G_0 \propto P$. We see that operating the diode in reverse bias makes sense, because the dark current, $-I_0$, is low compared to the photocurrent I_g.

We note right away that the photodiode, while very popular, is not without limitations. The diffusion process is typically slow, which limits the response time. Also, the optical absorption within the diffusion lengths L_p and L_n is very small, thus the conversion of optical power to photocarriers is limited.

13.1.2 Photovoltaic versus photoconductive mode

When used with zero-bias, the photodiode is said to operate in *photovoltaic mode*. Thus, with an open circuit, a voltage builds up, which acts as a forward bias (positive at the p-side). This is illustrated as the open-circuit voltage V_{oc} in figure 13.2. This photovoltaic effect is used in most solar cells. The *photoconductive mode* is achieved by applying a reverse bias to the junction. This mode of operation is most common as it produces low dark currents and shortens the response time. The response time is

Table 13.1. Bandgap energy and cutoff wavelength for various materials.

Material	E_g (eV)	λ_c (μm)
GaP	2.4	0.52
GaAs	1.4	0.93
Si	1.12	1.1
InGaAs	0.73, 0.59, 0.5	1.7
Ge	0.68	1.82
InAs	0.28	3.5
InSb	0.16	5.5
HgCdTe	Variable	Variable
PbSnTe	Variable	Variable

faster because the reverse bias increases the width of the depletion region and increases the electric field, which accelerates the carriers faster. At high reverse bias, the junction may break down, allowing for a multiplied photocurrent generation. In this mode, the diode exhibits an intrinsic *gain*, which increases the responsivity of the device. This is the *avalanche photodiode* mode, as discussed already and illustrated in figure 12.20.

13.1.3 Materials

Because the material bandgap controls the photon energy necessary to generate photocarriers, different materials operate at different wavelength spectra. Table 13.1 summarizes the spectral range of operation for several materials. Silicon is by far the most common material for photodiodes. Because of its higher energy bandgap compared to, say, germanium, thermal energy is less likely to generate carriers and, thus, Si-based photodiodes are less noisy.

13.1.4 Noise contributions

The p–n photodiode is subject to noise, like any other detector. The most important noise contributions are as follows.

Johnson noise

The *rms* Johnson current, which is due to the thermal fluctuations in the material, is (equation 10.49)

$$i_{rms} = \sqrt{\frac{4k_B T}{R}\Delta f}. \tag{13.6}$$

In equation (13.6), R is the *dynamic resistance* of the diode, defined as

$$R = \frac{dV}{dI} = \frac{1}{dI/dV}. \tag{13.7}$$

Using the I–V diode equation (equation (13.3)), $I = I_o(e^{eV/K_BT} - 1)$, we obtain readily

$$R = \frac{k_B T}{eI_0} e^{-eV/k_B T}. \tag{13.8}$$

Combining equations (13.6) and (13.8), we obtain

$$i_{rms} = \sqrt{4eI_0\Delta f e^{\frac{eV}{2k_BT}}}. \tag{13.9}$$

At zero bias, $i_{rms}\, \alpha \sqrt{I_0}$, where I_0 is the reverse saturation current, defined as (see equation (13.4))

$$I_0 = eA\left(\frac{n_p D_n}{L_n} + \frac{p_n D_p}{L_p}\right), \tag{13.10}$$

where e is the electronic charge, A the area of the junction, D_n, D_p the diffusion coefficients of electrons and holes, respectively $\left(D = \frac{k_B T}{e}\mu\right)$, n_p, p_n the minority carrier concentrations, L_n, L_p minority carrier diffusion lengths ($L = \sqrt{D\tau_{\text{lifetime}}}$). Overall, I_0 has a strong dependence on temperature, increasing at higher temperatures. It is expected that higher temperatures generate noisier signals. Thus, Johnson noise can be minimized by cooling the diode.

Shot noise

For a given bandwidth, Δf, the standard deviation of the current due to shot noise is (see equation (10.54))

$$i_{rms} = \sqrt{2e\bar{i}\Delta f}. \tag{13.11}$$

In equation (13.11), $\bar{i}$ is the total average current,

$$\bar{i} = i_{\text{sig}} + i_{\text{bkg}} + i_{\text{dark}}, \tag{13.12}$$

where i_{sig} is the average signal current, i_{bkg} the average background current, and i_{dark} the average dark current.

13.1.5 Figures of merit

Quantum efficiency

Photodiodes benefit from relatively high quantum efficiency, of 60–80% (see section 11.1 for the definition).

Responsivity
The *spectral responsivity* (see section 11.2) has a strong dependence on the material and its bandgap. The spectral responsivity of a silicon photodiode is shown in figure 13.3.

Noise-equivalent power
Noise-equivalent power (NEP) is essentially the minimum detectable optical power (recall section 11.6). In p–n photodiodes, NEP is governed by the dark current and by the photocurrent generated by stray electromagnetic waves that reach the junction.

Gain
Photodiodes do not provide intrinsic gain (see section 11.8) unless operated in the *avalanche mode*. Still, avalanche photodiodes have a gain of 10^2–10^3, which is much lower than that in photomultipliers (10^5–10^8).

Spatial and temporal sampling
Temporal sampling (see section 11.10) is defined by the response time of the device. The EHPs generated in the junction undergo a finite transit time through the material, which tends to limit the response time. Additionally, the resistance and capacitance of the junction and deriving circuit contribute a time constant $\tau = RC$. This quantity stretches the impulse response of the device and ultimately limits the bandwidth of the modulation signal.

Spatially, common p–n photodiodes operate as a 'single point' detector. More recently, photodiode arrays, in which 100–1000 photodiodes form a 1D array, have become available. Such an arrayed device allows for 1D spatial sampling and high-speed parallel data readout.

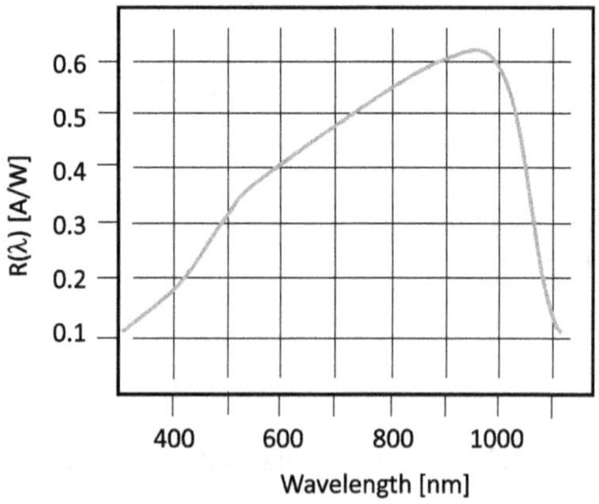

Figure 13.3. Spectral responsivity of silicon.

13.2 Photoconductive detectors

13.2.1 Photoconductivity

Photoconductive detectors operate by recording changes in the electrical conductivity induced by the incident light. We have already discussed how a p–n photodiode can operate as a photoconductive detector (section 13.1.2). However, such a detector can be realized by a single piece of semiconductor exposed to incident light (see figure 13.4).

The total current density through the material is

$$\begin{aligned} j &= \sigma E \\ &= j_n + j_p \\ &= (e\mu_n n + e\mu_p p)E, \end{aligned} \tag{13.13}$$

where e is the electronic charge, $\mu_{n,p}$ are the mobilities, and n and p are the concentrations of the electrons and holes, respectively. Thus, the conductivity is

$$\sigma = e(\mu_n n + \mu_p p). \tag{13.14}$$

The conductivity can be written as the sum of the *dark* conductivity, σ_0, and the conductivity charge in the presence of light, σ_1,

$$\begin{aligned} \sigma &= \sigma_0 + \sigma_1. \\ &= e(\mu_n n_0 + \mu_p p_0) + e(\mu_n n_1 + \mu_p p_1). \end{aligned} \tag{13.15}$$

In equation (13.15), n_0, p_0 are the dark carrier concentration, while n_1, p_1 are the charge in concentrations due to light. The total current in the circuit also consists of the 'light' and 'dark' components

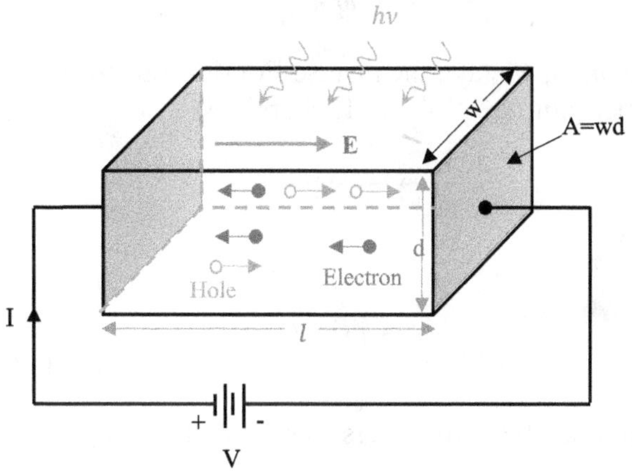

Figure 13.4. A photoconductor made of a p-doped semiconductor, under bias V.

$$\begin{aligned} I &= jA \\ &= \sigma EA \\ &= \sigma \frac{AV}{l} \\ &= I_0 + I_1. \end{aligned} \tag{13.16}$$

In equation (13.16), A is the transverse area and l is the length of the material (see figure 13.4). In the presence of light, photocarriers are generated (n_1, p_1) which lead to an increase in current

$$I_1 = e(\mu_e n_1 + \mu_p p_1)\frac{A}{l}V. \tag{13.17}$$

Thus, I_1 is the measurable quantity that reports on the incident optical power, P. The responsivity of this photoconductive detector is $R = I/P$.

13.2.2 Response time

Let us consider that the light is delivered as an infinitely short pulse, thus, the carrier generation is $g(t) = g_0\delta(t)$. The rate equation for electrons is

$$\frac{\partial}{\partial t}n_1(t) = g_0\delta(t) - \frac{n_1(t)}{\tau_n}, \tag{13.18}$$

where τ_n is the recombination time for electrons. Taking the Fourier transform of equation (13.18), we can immediately solve for $n_1(\omega)$ (see Volume 1, chapter 4, for the differentiation properties of the Fourier transform)

$$n_1(\omega)\left[-i\omega + \frac{1}{\tau_n}\right] = g_0 \tag{13.19a}$$

$$n_1(\omega) = \frac{ig_0}{\omega + \frac{i}{\tau_n}}. \tag{13.19b}$$

In order to obtain $n_1(t)$, we take the inverse Fourier transform of equation (13.19*b*) and use the Fourier pair (see Volume 1, chapter 4)

$$\frac{1}{\omega} \leftrightarrow -i\ \mathrm{sign}(t) \tag{13.20a}$$

and the *shift theorem*

$$\frac{1}{\omega + \frac{i}{\tau_n}} \leftrightarrow i\ \mathrm{sign}(t)e^{-\frac{t}{\tau}}. \tag{13.20b}$$

Thus, the injection of photocarriers due to a light impulse has the form (see figure 13.5)

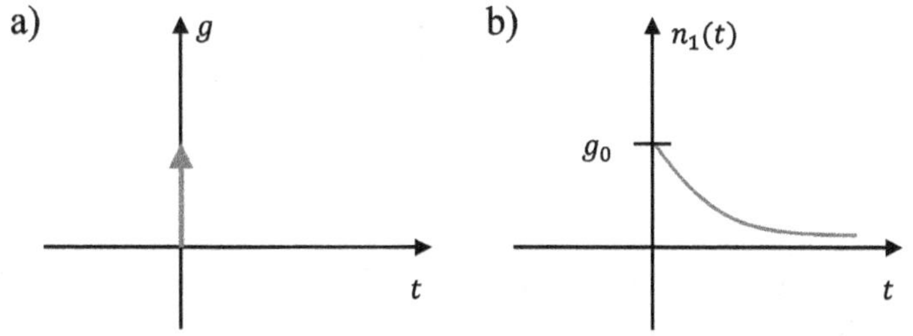

Figure 13.5. Generation rate for an impulse of light. (a) Infinitely short incident pulse. (b) Carrier concentration time dependence.

$$n_1(t) = g_0 e^{-\frac{t}{\tau_n}},\ t \geqslant 0. \tag{13.21}$$

From equation (13.17) we see that the change in the measurable current is proportional to n_1. Thus, from equation (13.21) we conclude that the photodetector has a finite response time, of the order of the recombination time τ_n. Ultimately, τ_n limits the maximum optical modulation frequency that the photodetector can measure.

It is left as an exercise to calculate $n_1(t)$ for other time dependences of the input light: step function, rectangular function, and constant (see problem 13.2).

13.3 Photoemission detectors

13.3.1 Photocathodes

Photoemissive detectors operate on the principle of the photoelectric effect: the emission of an electron from the surface of a material following the absorption of incident photons. Famously, the explanation of the photoelectric effect was put forward by Einstein in 1905, for which he received the Nobel prize [5].

Let us consider a metal surface illuminated by monochromatic radiation of optical frequency, ν. The electrons emitted from the material have a kinetic energy equal to

$$W = h\nu - \varphi. \tag{13.22}$$

This kinetic energy is the difference between the photon energy, $h\nu$, and the binding energy of the electron, φ, known as the *work function*. At $T = 0$ K, the highest energy level that the electrons can occupy the metal is the Fermi level (see chapter 12). Thus, the work function is the energy needed to take the electron from the Fermi level to the *vacuum level*, defined as the energy of the electron far away from the metal surface, with zero kinetic energy (figure 13.6).

Note that, at room temperature, there are electrons with energies above the Fermi level, which will borrow the value of φ. Typical values for φ are of the order of several eV. Cesium (Cs) has one of the lowest work functions, $\varphi = 2$ eV, which corresponds to a wavelength $\lambda = 620$ nm.

The photoelectric effect can also take place in semiconductors. The interpretation of this phenomenon must take into consideration the bandgap structure associated

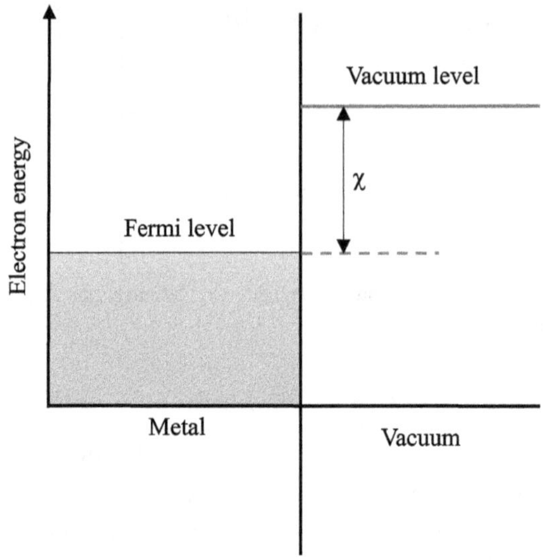

Figure 13.6. Photoelectric effect: energy levels in a metal.

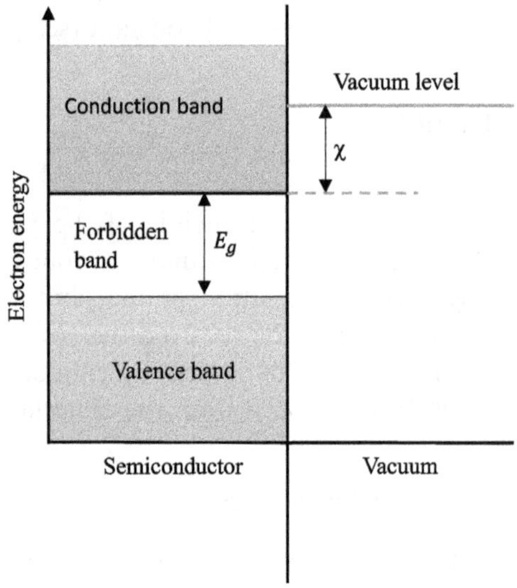

Figure 13.7. Photoelectric effect: energy levels in a semiconductor.

with a semiconductor (figure 13.7). The energy difference between the vacuum level and the bottom of the conduction band is called *electron affinity*, χ. Thus the minimum photon energy required to emit an electron for the valence band is (figure 13.7)

$$h\nu = E_g + \chi. \tag{13.23}$$

The most common photoemissive detectors are the *vacuum photodiode* and the *photon multiplier*, as discussed next.

13.3.2 Photodiodes

The vacuum photodiode is the simplest photoemissive detector. Figure 13.8 depicts a vacuum photodiode, which consists of a photocathode (the electron emitter) and an anode, separated by a distance *d*. The diode is under forward bias, *V*, such that an electron ejected from the cathode is accelerated toward the anode by a force

$$m\ddot{x}(t) = eE = \frac{eV}{d}. \tag{13.24}$$

Assuming $x(t = 0) = 0$ and $\dot{x}(t = 0) = 0$ (i.e., zero initial velocity), we solve equation (13.24) and obtain

$$x(t) = \frac{eV}{2md}t^2. \tag{13.25}$$

Setting $x(\tau) = d$, we obtain the time necessary for the electron to cross the distance between the electrodes

$$\tau = d\sqrt{\frac{2m}{eV}}. \tag{13.26}$$

Note that τ defines the finite time response of the photodiode. A higher voltage V decreases the transit time τ and, thus, the detector response time, but increases the dark current due to charges that make it to the anode in the absence of light. We can conclude that the responsivity $R(f)$ has a cut-off frequency $f_e \simeq 1/2\tau$.

13.3.3 Photomultipliers

A photomultiplier consists of cascading multiple diodes and amplifying the electron emission. Figure 13.9 describes a photomultiplier. The emitted *primary* electrons are

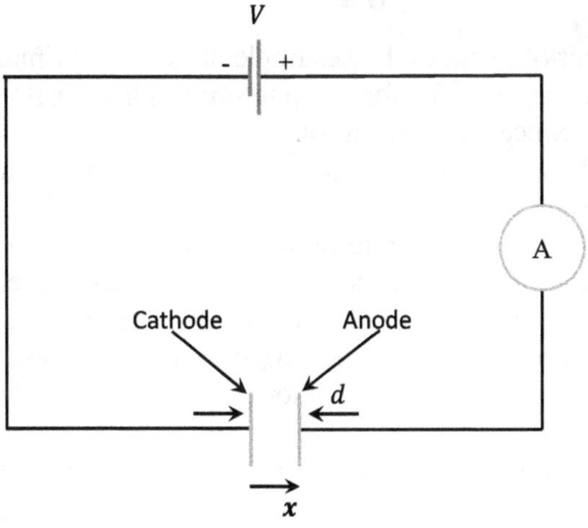

Figure 13.8. Vacuum photodiode.

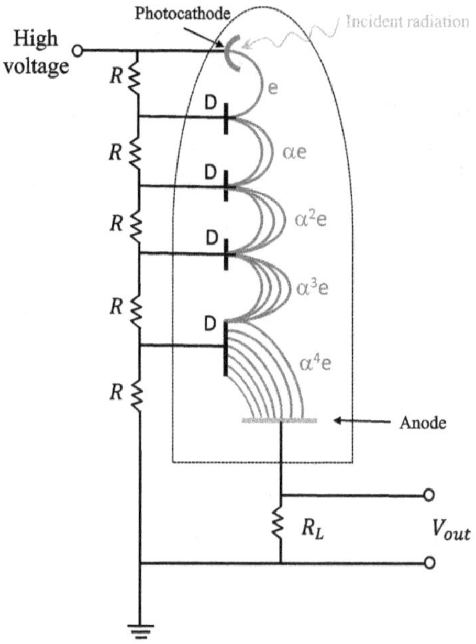

Figure 13.9. Schematic of a photomultiplier: R is the resistor, α the secondary emission ratio, D the dynode, R_L the load resistor, and V_{out} the output voltage.

accelerated to the anode and generate *secondary* electrons. This anode acts as a new photocathode because the electrode is kept at a higher voltage. These electrodes are called *dynodes*. Every time an electrode strikes a dynode, on average α ($\alpha > 1$) electrons are emitted, which leads to an amplification process. The total gain of the photomultiplier, G, is the ratio of the electrons finally reaching the anode for each electron departing at the cathode. If we set the *secondary emission ratio* as α, then the photomultiplier gain is

$$G = \alpha^N. \tag{13.27}$$

Here N is the number of dynodes. For example, if for each primary electron we have three secondary ones, $\alpha = 3$, for a photomultiplier that has ten dynodes, $G = 3^{10} = 6 \times 10^4$. Since small variations in voltage (and, thus, in α) can lead to large variations in the current at the anode, photomultipliers use extremely stable power supplies.

The main noise contributors in photomultipliers are the *shot noise* due to the dark current, and noise due to the amplification process. Dark current, produced in the absence of light, is due to the thermal emission of electrons from the photocathode. Cooling the photocathode decreases the dark current. Other sources of dark current include the potential presence of residual ions in the tube and electron emission due to cosmic radiation.

For incident radiation of frequency ν and power P, the current at the cathode is

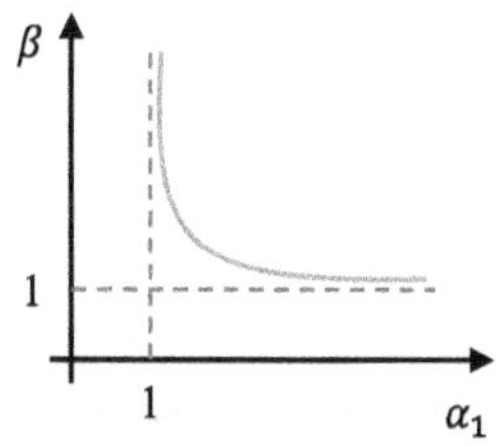

Figure 13.10. SNR decrease (β) vs the secondary emission ratio for the first amplification stage (α_1).

$$i_C = i_D + \frac{\eta e P}{h\nu}, \tag{13.28}$$

where i_D is the dark current and η is the quantum efficiency. The shot noise *rms* current is

$$i_C^{rms} = \sqrt{2ei_C\Delta f}. \tag{13.29}$$

The current at the anode is the one at the cathode amplified by the gain

$$i_A = Gi_C. \tag{13.30}$$

Thus, the *rms* shot noise at the anode is

$$\begin{aligned} i_A^{rms} &= \sqrt{2Gei_c\Delta f} \\ &= \sqrt{2ei_A\Delta f}. \end{aligned} \tag{13.31}$$

In real experimental situations, the secondary emission ratio is a stochastic process. For a Poisson distribution (see section 9.4.2), the variance equals the average, $\sigma_\alpha^2 = <\alpha>$. In this case it has been shown that the current signal-to-noise ratio (SNR) decreases by the factor [1]

$$\beta = \left(\frac{\alpha_1}{\alpha_1 - 1}\right)^{1/2}, \tag{13.32}$$

where α_1 is the secondary emission ratio for the first amplification stage. The first stage of amplification is the most significant source of noise because it is amplified the most by the subsequent stages. Figure 13.10 shows the degradation of SNR with respect to α_1.

The shot noise result can be modified to include the additional noise due to amplification (equation (13.32)),

$$i_A^{rms} = \beta\sqrt{2Gei_A\Delta f}. \tag{13.33}$$

13.4 Problems

1. A p–n junction under voltage bias V_1 has a dynamic resistance R_1. What is the difference between a new voltage V_2 and V_1, that is, $V_2–V_1$, that will reduce the resistance by half ?
2. A photodetector is characterized by an electronic recombination time τ_n. Calculate the electron concentration versus time if the EHP generation rate has the form:
   a) $g(t) = g_o\Pi\left(\frac{t}{T}\right)$
   b) $g(t) = g_0\Gamma(t)$
   c) $g(t) = g_0(1 + \cos \Omega t)$.
3. Calculate the conductivity relative change due to photoelectrons between time $t = 0$ and $t = \tau_n/2$, in all three cases of problem 2.
4. What is the longest wavelength of a photon that can create a photoelectron from a material with a work function of 1 eV, 10 eV, 100 eV?
5. In a vacuum photodiode, what is the voltage between the cathode and the anode that results in a transit time of 1 μs, if the distance between electrodes is 1 cm?
6. Consider a photomultiplier with a secondary emission ratio $\alpha = 5$, and consisting of eight dynodes. Assuming a negligible dark current, a quantum efficiency $\eta = 0.9$, an incident optical power $P = 1$ nW at $\lambda = 500$ nm, and a bandwidth $\Delta f = 150$ kHz, calculate the following:
   a) the shot noise current at the cathode;
   b) the shot noise current at the anode.

References

[1] Boyd R W 1983 *Radiometry and the Detection of Optical Radiation* (Wiley Series in Pure and Applied Optics) (New York: Wiley), vii p 254

[2] Dereniak E L and Boreman G D 1996 *Infrared Detectors and Systems* vol 306 (New York: Wiley)

[3] Chuang S L 1995 *Physics of Optoelectronic Devices* (Wiley Series in Pure and Applied Optics) (New York: Wiley), xv p 717

[4] Streetman B G and Banerjee S 2006 *Solid State Electronic Devices* 6th edn (Prentice Hall Series in Solid State Physical Electronics) (Upper Saddle River, NJ: Pearson/Prentice Hall), xviii p 581

[5] Stachel J J (ed) 1998 *Einstein's Miraculous Year: Five Papers that Changed the Face of Physics* (Princeton, NJ: Princeton University Press)

IOP Publishing

Principles of Biophotonics, Volume 2

Light emission, detection, and statistics

Gabriel Popescu

Chapter 14

Thermal detectors

14.1 Principle of photothermal detection

Thermal detectors operate on the principle of the photothermal effect: the change in temperature produced by an active element following the absorption of light [1–4]. Because absorption and emission of radiation are reverse of each other (Kirchhoff's law), a thermal detector is also an emitter of thermal radiation.

Let us consider the thermal detector depicted in figure 14.1. The active element of the detector is in contact with a *heat sink* kept at constant temperature T_0. The incident light causes a slight temperature increase in the element, $T_1 (T_1 \ll T_0)$. The heat is transferred from the element to the heat sink. The heat flow to the sink per unit time, that is, the power dissipated into the sink, P_s, is

$$P_s = \Sigma T_1, \tag{14.1}$$

where Σ is the thermal conductance (in W/k), analog to the electric conductance. The detector is characterized by a heat capacity C, defined as

$$C = \frac{dQ}{dT_1}, \tag{14.2}$$

which represents the amount of heat, dQ, absorbed by the detector that will raise its temperature by dT_1, $[C] = J/k$.

Let us consider the time varying optical power $P(t)$ falling onto the detector. To balance the energy flow in the system, $dQ/dt = 0$, the following equation holds

$$\frac{d}{t}(Q_{\text{stored}} + Q_{\text{conduct}}) = \frac{dQ_{\text{absorbed}}}{dt} \tag{14.3a}$$

$$C\frac{dT_1(t)}{dt} + \Sigma T_1(t) = \varepsilon P(t), \tag{14.3b}$$

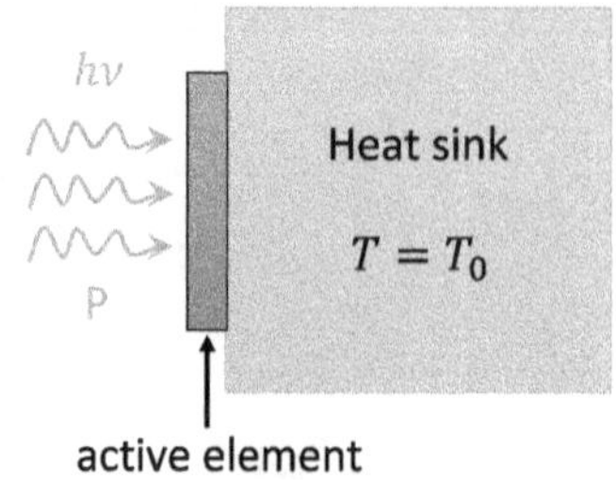

Figure 14.1. Thermal detector under light illumination of frequency ν and power P.

where ε is the fraction of the incident power absorbed by the detector. In equations (14.3a and b), the first term on the LHS is the power stored due to the heat capacity of the detector, the second term on the LHS is the power transferred to the sink, and the RHS is the power absorbed from the light.

Let us assume, $P(t) = P_0\Gamma(t)$, meaning that the optical power turns on at $t = 0$, that is, it has a step function shape. To solve for $T_1(t)$, we take the Laplace transform of equation (14.3b) (see Volume 1, chapter 11)

$$\left(sC + \sum\right)T_1(s) = \frac{\varepsilon P_0}{s}, \tag{14.4a}$$

thus, the solution in the Laplace domain is

$$T_1(s) = \frac{\varepsilon P_0}{C} \cdot \frac{1}{s\left(s + \sum/C\right)}. \tag{14.4b}$$

To obtain equation (14.4a), we used the differentiation theorem, $d/dt \leftrightarrow s$, and $\Gamma(t) \leftrightarrow 1/s$. In order to obtain the time domain solution we take the inverse Laplace transform of equation (14.4b), which becomes easy once we arrange $T_1(s)$ using the partial fraction decomposition (see Volume 1, chapter 11).

$$T_1(s) = \frac{\varepsilon P_0}{\Sigma}\left[\frac{1}{s} - \frac{1}{s + \Sigma/C}\right]. \tag{14.5}$$

Thus, the time domain solution is obtained by invoking the shift theorem, $1/(s + \sum/c) \leftrightarrow e^{-\frac{\Sigma}{c}t}\Gamma(t)$, namely

$$T_1(t) = \frac{\varepsilon P_0}{\Sigma}\left(1 - e^{-\frac{\Sigma}{C}t}\right)\Gamma(t). \tag{14.6}$$

Figure 14.2 illustrates the time dependence of T_1.

Clearly, the temperature increase has a characteristic, response time $\tau = C/\Sigma$. Thus, the temporal responsivity of the detector will have a cut-off at $f_{max} = \frac{1}{2\tau}$. A small heat capacity and large conductivity are desirable for fast response.

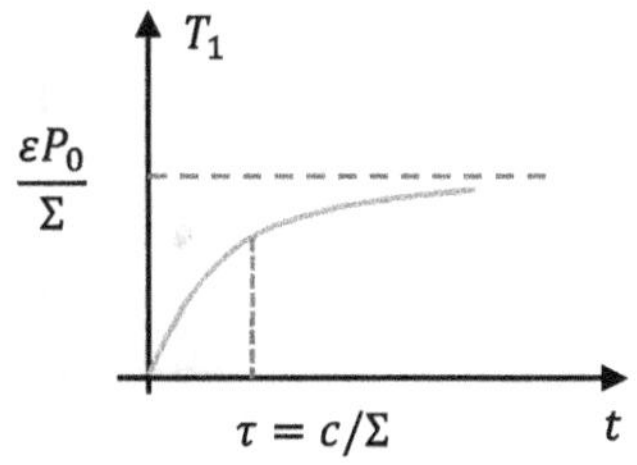

Figure 14.2. Temperature change for an incident optical power in the form of temporal step function at the origin.

14.2 Noise in thermal detectors

The main source of noise in a thermal detector is due to thermal fluctuations. In the absence of incident radiation, the mean-squared fluctuations in T_1 is (see p 338 in [5])

$$\overline{\Delta T_1^2} = \frac{k_B T_0^2}{2}. \tag{14.7}$$

Equation (14.7) is sometimes referred to as the Fowler–Einstein equation. The power spectrum of the temperature fluctuations has a form analog to the Johnson noise (section 10.3.1)

$$\overline{\Delta T_1^2}(f) = \frac{4k_B T_0^2/\Sigma}{1 + (2\pi f \tau)^2}. \tag{14.8}$$

In a narrow band of frequencies, Δf, the *rms* fluctuations are

$$\Delta T_1^{rms} = \sqrt{\overline{\Delta T_1^2}(f)\Delta f}. \tag{14.9}$$

The optical signal power required to produce a signal equal to ΔT_1^{rms}, that is, the *noise equivalent power* (NEP) is

$$\text{NEP} = \frac{\Delta T_1^{rms}}{|R_T(f)|}, \tag{14.10}$$

where $R_T(f)$ is the frequency responsivity. In order to find $R_T(f)$, we take the Fourier transform of equation (14.6),

$$T_1(f) = \frac{\varepsilon P_0}{\Sigma}\left[\delta(f) - \frac{1}{1 + i(C/\Sigma)f}\right]. \tag{14.11}$$

Filtering out the DC contribution, $\delta(f)$, the magnitude of $T_1(f)$ is

$$|T_1(f)| = \frac{\varepsilon P_0}{\Sigma}\frac{1}{(1 + 4\pi^2 f^2 \tau^2)^{1/2}}. \tag{14.12}$$

Thus, the frequency responsivity, R_T, $[R_T] = \frac{K}{W}$, is

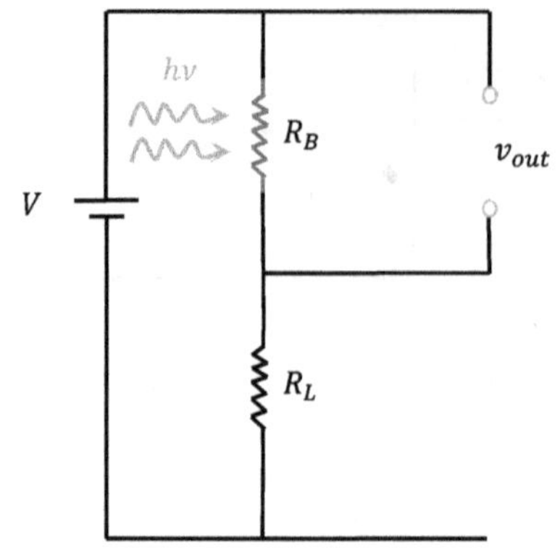

Figure 14.3. Bolometer circuit: R_B is the bolometer resister, R_L is the load resister.

$$\begin{aligned} R_T(f) &= \frac{|T_1(f)|}{P_0} \\ &= \frac{\varepsilon / \sum}{(1 + 4\pi^2 f^2 \tau^2)^{1/2}}. \end{aligned} \tag{14.13}$$

Finally, the NEP is obtained by plugging equation (14.13) into equation (14.10),

$$\text{NEP} = \frac{1}{2}\sqrt{\sum 4k_B T_0^2 \Delta f}. \tag{14.14}$$

Equation (14.14) indicates that one can achieve a lower NEP, which is desirable, by increasing the emissivity ε and by lowering the thermal conductivity $\sum$ and the sink temperature T_0.

14.3 Bolometers

A bolometer is a thermal detector that detects temperature changes via fluctuations in its resistance. Figure 14.3 illustrates a simple bolometer circuit. Let us consider radiation of power P incident to the bolometer resistor R_B. It is useful in practice to use an identical *load* resistor R_L, which is shielded from the incident radiation. Using two identical resistors automatically cancels out fluctuations in the voltage V and ambient temperature fluctuations. The change in resistance, ΔR, with temperature, ΔT, is

$$\Delta R = \alpha R_B \Delta T, \tag{14.15}$$

where α is the temperature coefficient of resistance, defined as

$$\alpha = \frac{1}{R}\frac{dR}{dT}. \tag{14.16}$$

This change in resistance leads to a change in the output voltage

$$\begin{aligned} \nu &= i\Delta R \\ &= i\alpha R_B \Delta T, \end{aligned} \tag{14.17}$$

where the current i is inferred via Ohm's law

$$i = \frac{V}{R_L + R_B}. \tag{14.18}$$

The total voltage noise is the sum of the contributions from the Johnson noise and temperature noise, namely

$$\overline{\nu^2} = 4k_BTR\Delta f + i^2\alpha^2R_B^2\overline{(\Delta T)^2}\Delta f, \tag{14.19}$$

or, using equation (14.8) for frequencies $f \ll \frac{1}{2\pi\tau}$,

$$\overline{\nu^2} = 4k_BTR\Delta f + i^2\alpha^2R_B^2\frac{4k_BT^2}{\Sigma}\Delta f. \tag{14.20}$$

In equation (14.20), R denotes the combination of R_B and R_L. For the series connection in figure 14.3, $R = R_B + R_L = 2R_B$, assuming identical resistors.

The bolometer is considered ideal if the Johnson noise is negligible compared to the temperature noise,

$$\frac{R\Sigma}{i^2\alpha^2R_B^2T} \ll 1. \tag{14.21}$$

Under these ideal conditions, the expression for NEP is given by equation (14.14). It is left as an exercise to calculate the NEP when the noise is Johnson dominated (problem 14.4).

14.4 Pyroelectric detectors

Pyroelectric detectors ('pyro' means 'fire' in Greek) are crystals that exhibit permanent electric dipole moments, which change with temperature. As the temperature of the detector varies, the charges in the crystal are displaced, thus creating an electric current. Let us consider the pyroelectric detector in figure 14.4.

The magnitude of the current induced in the crystal due to the temperature change is

$$i(t) = pA\frac{dT(t)}{dt}. \tag{14.22}$$

In equation (14.22), p is the *pyroelectric coefficient*, $[p]$ = Coulomb/m^2 K and A is the area of the electrode. Compared to a bolometer, the pyroelectric responds to the time derivative of the temperature change, not the temperature change itself.

Using the differentiation theorem (Volume 1, chapter 4), we Fourier transform equation (14.22) to obtain

$$i(f) = i2\pi fpA\Delta T(f). \tag{14.23}$$

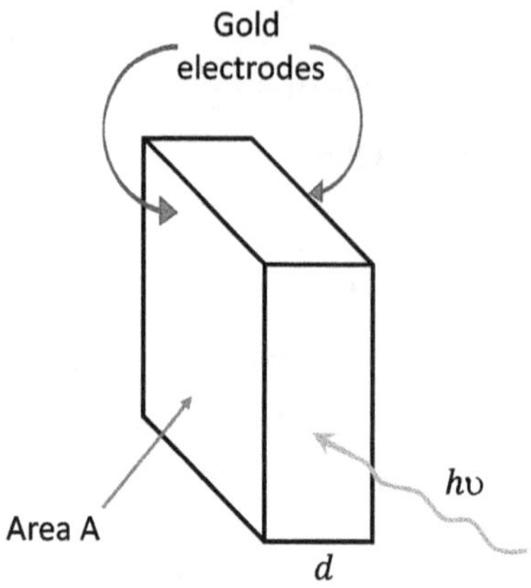

Figure 14.4. Pyroelectric detector.

Thus, using equation (11.16), we can infer immediately the current responsivity for pyroelectrics ($\tau = C/\Sigma$)

$$|R_i(f)| = \frac{\varepsilon p A}{\Sigma}\frac{2\pi f}{\sqrt{1+(2\pi f\tau)^2}}. \tag{14.24}$$

Johnson noise is the primary noise source,

$$i_{rms} = \sqrt{\frac{4k_BT}{R}\Delta f}. \tag{14.25}$$

The NEP has the form

$$\mathrm{NEP} = \frac{\sqrt{4k_BT\Delta f/R}}{pA\varepsilon/\sum\tau}. \tag{14.26}$$

Equation (14.26) indicates that the NEP can be lowered by increasing the value of R and a large pyroelectric coefficient. However, as shown in figure 14.5, increasing the resistance tends to reduce the frequency response.

14.5 Problems

1. Let us consider a thermal detector, where the active element has a heat capacity C and is connected to a heat sink through a thermal conductance $\sum$. If the incident light is a rectangular pulse of width T, calculate the time-dependent temperature change, if 50% of the light is absorbed.
2. What is the thermal conductance of a thermal detector at room temperature, if a bandwidth of 100 Hz exhibits an NEP = 1 μW?
3. For a bolometer using a load resistor in parallel and equal to its own resistance, what is the thermal conductivity for which the Johnson noise equals temperature noise ($T = 300$ K, $R_B = 100$ kΩ, $\alpha = 10^{-3}$ K^{-1})?

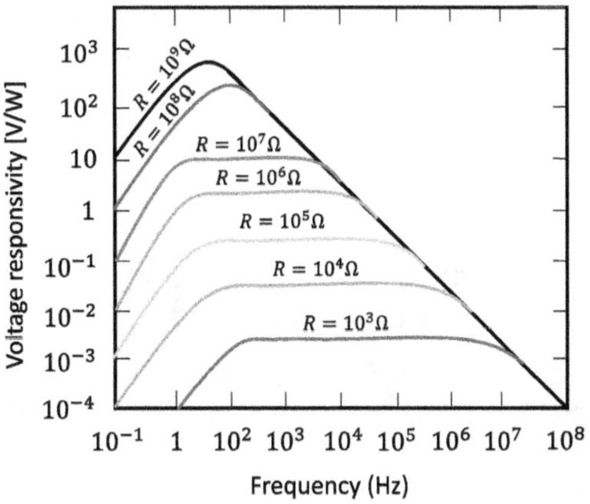

Figure 14.5. Frequency response of typical pyroelectrics for several values of the load resistance.

4. Find the expression for the NEP of a bolometer in a Johnson-dominated regime.

References

[1] Ahmed S N 2007 *Physics and Engineering of Radiation Detection* 1st edn (Amsterdam: Academic), xxiv p 764
[2] Boyd R W 1983 *Radiometry and the Detection of Optical Radiation* (Wiley Series in Pure and Applied Optics) (New York: Wiley), vii p 254
[3] Dereniak E L and Boreman G D 1996 *Infrared Detectors and Systems* vol 306 (New York: Wiley)
[4] Bass M 1995 *Optical Society of America: Handbook of Optics* 2nd edn (New York: McGraw-Hill)
[5] Landau L D, Lifshitz E M and Pitaevskii L P 1980 *Statistical Physics: Course of Theoretical Physics* (Oxford: Pergamon)

IOP Publishing

Principles of Biophotonics, Volume 2
Light emission, detection, and statistics
Gabriel Popescu

Chapter 15

Statistics of optical fields

15.1 Optical fields as random variables

All optical fields encountered in practice are subject to statistical *uncertainty*. The *random*, *unpredictable* fluctuations in both space and time of optical fields are rooted in respective fluctuations of the sources (both primary and secondary). The discipline that studies these fluctuations is known as *coherence theory* or *statistical optics* [1, 2]. The coherence properties of optical fields are crucial for the outcomes of experiments. Whenever we measure a superposition of fields (e.g. in interferometry, imaging, etc) the result of the statistical average performed by the detection process is strongly dependent on the coherence properties of the light. It is significant that ½ of the 2005 Nobel Prize in Physics was awarded to Roy Glauber 'for his contribution to the quantum theory of optical coherence'. For a selection of Glauber's seminal papers, see [3].

The origin of the stochastic (random) fluctuations in the electric field is found in the emission process itself. For example, a thermal source, such as a bulb filament or the surface of the Sun, emits light in a manner that cannot be predicted with certainty. In other words, we cannot find a function $f(\mathbf{r}, t)$ that prescribes the field at each point in space and each moment in time. Instead, we describe the source as emitting a random signal, $s(\mathbf{r}, t)$ (figure 15.1). We can gain knowledge about the random field only by repetitive measurements and subsequent averaging of the results. This type of averaging over many manifestations (*realizations*) of a certain random variable is called *ensemble averaging*. The importance of ensemble averaging has been stressed many times by both Glauber and Wolf. For example, on page 29 of [3], Glauber mentions 'It is important to remember that this average is an ensemble average. To measure it, we must in principle repeat the experiment many times by using the same procedure for preparing the field over and over again. That may not be a very convenient procedure to carry out experimentally but it is the only one which represents the precise meaning of our calculation'.

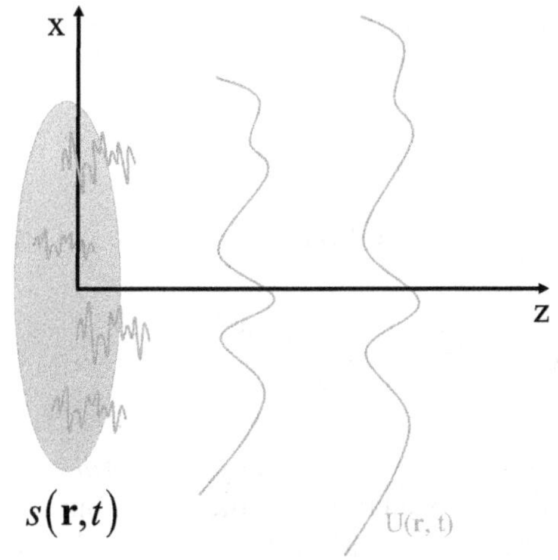

Figure 15.1. An arbitrary source emitting a random signal $s(\mathbf{r}, t)$, which is the result of uncorrelated point sources. $U(\mathbf{r}, t)$ is the field at a certain distance from the source (assumed to be scalar).

An arbitrary source can be thought of as consisting of many point sources, each emitting fields of certain correlations *in time* and together exhibiting certain correlations *in space.* However, as we will discuss later, it is important to realize that the statistical properties of the optical fields change with propagation. How stochastic fields are transformed by free-space propagation and various optical components can be understood through the formalism of linear systems with random inputs (Volume 1, chapter 9).

15.2 Spatiotemporal correlation function

Optical coherence is a manifestation of the *field statistical similarities* in space and time and coherence theory is the discipline that mathematically describes these similarities [1]. A *deterministic*, fully *predictable* field distribution in both time and space is the *monochromatic plane wave*, which, of course, is only a mathematical construct, impossible to obtain in practice due to the uncertainty relation (recall Volume 1, chapter 8).

Perhaps a starting point in understanding the physical meaning of a *statistical optical field* is to ask the question: what is the *effective (average) temporal* sinusoid, that is, $\langle e^{-i\omega t}\rangle_\omega$, for a broadband field? Similarly, what is the *average spatial* sinusoid, that is, $\langle e^{i\mathbf{k}\cdot\mathbf{r}}\rangle_\mathbf{k}$? Here we use the sign convention whereby a monochromatic plane wave is described by $e^{-i(\omega t-\mathbf{k}\cdot\mathbf{r})}$ (see Volume 1, section 3.2). These two averages can be performed using the *probability densities* associated with the temporal and spatial frequencies, $S(\omega)$ and $P(\mathbf{k})$, respectively, which are normalized to satisfy $\int S(\omega)d\omega = 1$ and $\int P(\mathbf{k})d^3\mathbf{k} = 1$. Thus, $S(\omega)d\omega$ is the probability of having frequencies in the range $(\omega, \omega + d\omega)$ in our field, or the fraction of the total power contained in the vicinity of frequency ω. Similarly, $P(\boldsymbol{k})d^3\mathbf{k}$ is the probability of having spatial frequencies in the infinitesimal volume around $\mathbf{k}$, or the fraction of the total power contained around spatial frequency $\mathbf{k}$. Up to a normalization factor,

S and P are the *temporal and spatial power spectra* associated with the fields. Thus, the two 'effective sinusoids' can be expressed as *ensemble averages*, using $S(\omega)$ and $P(\boldsymbol{k})$ as weighting functions,

$$\begin{aligned}\langle e^{-i\omega t}\rangle_\omega &= \int S(\omega)e^{-i\omega t}\, d\omega \\ &= \Gamma(t)\end{aligned} \tag{15.1a}$$

$$\begin{aligned}\langle e^{i\mathbf{k}\cdot\mathbf{r}}\rangle_\mathbf{k} &= \int P(\mathbf{k})e^{i\mathbf{k}\cdot\mathbf{r}}\, d^3\mathbf{k} \\ &= W(\mathbf{r}).\end{aligned} \tag{15.1b}$$

Equations (15.1*a*–*b*) establish that the average *temporal sinusoid* for a broadband field equals its temporal autocorrelation, denoted by Γ. Similarly, the average *spatial sinusoid* for an inhomogeneous field equals its spatial autocorrelation, denoted by W.

Besides the basic scientific interest in describing the statistical properties of optical fields, coherence theory can make predictions of experimental relevance. The general problem can be formulated as follows (figure 15.1): given the optical field distribution $U(\mathbf{r}, t)$ that varies randomly in space and time, over what *spatiotemporal domain* does the field preserve significant correlations? Experimentally, this question translates into: combining the field $U(\mathbf{r}, t)$ with a replica of itself shifted in both time and space, $U(\mathbf{r} + \boldsymbol{\rho}, t + \tau)$, on average, how large can $\boldsymbol{\rho}$ and τ be and still observe 'significant' interference?

Intuitively, we expect that monochromatic fields exhibit (infinitely) broad temporal correlations, while plane waves are expected to manifest infinite spatial correlations. This is so because regardless of how much we shift a monochromatic field in time or a plane wave in space, they remain perfectly correlated with their unshifted replicas. On the other hand, it is difficult to picture temporal correlations decaying over timescales that are shorter than an optical period and spatial correlations that decay over spatial scales smaller than the optical wavelength. In the following, we provide a quantitative description of the spatiotemporal correlations.

The statistical behavior of an optical field U, assumed to be scalar, can be mathematically captured quite generally via its *spatiotemporal correlation function*

$$\Lambda(\mathbf{r}_1, \mathbf{r}_2; t_1, t_2) = \langle U(\mathbf{r}_1, t_1)U^*(\mathbf{r}_2, t_2)\rangle, \tag{15.2}$$

where the angular brackets denote the *ensemble average*. In essence, this autocorrelation function quantifies how similar the field is with respect to a shifted version of itself, in time or space. This ensemble average procedure requires the collection of a large number of samples. For example, to measure the average of the field in time and space, one needs to acquire 'snapshots' of the signal, $U(t)_m$ and $U(\mathbf{r})_n$, respectively, and average them together. Thus, the temporal and spatial averages of the field can be obtained as

$$\langle U(t)\rangle = \frac{1}{M}\sum_{m=1}^{M} U(t)_m$$

$$\langle U(\mathbf{r})\rangle = \frac{1}{N}\sum_{n=1}^{N} U(\mathbf{r})_n.$$

To obtain the spatiotemporal correlation function defined in equation (15.2), one needs to average over the product of the samples in both time and space (a total of four summations). Thus, performing ensemble averages is often difficult, as pointed out by Glauber [3] and mentioned in section 15.1. However, a simplifying assumption, described next, allows us to compute averages from a single, sufficiently long, sample.

15.3 Ergodic hypothesis

A random process is considered *ergodic* if its statistical properties can be inferred from a single (sufficiently long) random sample of the process (see section 13.1 in [4] and Volume 1, section 9.4). In other words, acquiring a long enough spatial or temporal sequence of the random variable of interest, say, the value of the field U, captures all the possible realizations of the variable. The statistics of the process are captured by this single sample of the process. Thus, the ensemble average can be replaced by the time (or space) average. The spatiotemporal correlation function (equation (15.2)) can be re-written as follows:

$$\Lambda(\mathbf{r}_1, \mathbf{r}_2; t_1, t_2) = \langle U(\mathbf{r}_1, t_1)U^*(\mathbf{r}_2, t_2)\rangle_{\mathbf{r},t}. \tag{15.3}$$

The average <> is now performed both *temporally and spatially*, as indicated by the subscripts **r** and t. Because common detector arrays capture the spatial intensity distributions in 2D only, we can first discuss the 2D spatial dependence, $\mathbf{r} = (x, y)$. The time and space averages are defined in the usual sense, respectively, as

$$\begin{aligned}\langle U(\mathbf{r}_1, t_1)\cdot U^*(\mathbf{r}_2, t_2)\rangle_t &= \lim_{T\to\infty}\frac{1}{T^2}\int_{-T/2}^{T/2}\int_{-T/2}^{T/2} U(\mathbf{r}_1, t_1)U^*(\mathbf{r}_2, t_2)dt_1\,dt_2\\ \langle U(\mathbf{r}_1, t_1)\cdot U^*(\mathbf{r}_2, t_2)\rangle_{\mathbf{r}} &= \lim_{A\to\infty}\frac{1}{A^2}\int_A\int_A U(\mathbf{r}_1, t_1)U^*(\mathbf{r}_2, t_2)d^2\mathbf{r}_1\,d^2\mathbf{r}_2,\end{aligned} \tag{15.4}$$

where A is the area of interest, defined by the spatial integration domain. Note that these averages are now very practical to perform experimentally. From a single time sequence or spatial distribution recorded by the detector, we can easily measure field statistics. Further simplifications occur if we can make assumptions about the spatiotemporal behavior of the statistical quantities of interest.

15.4 Stationarity and statistical homogeneity

Often, in practice we deal with fields that are both *stationary* (in time) and *statistically homogeneous* (in space), as discussed for generic stochastic signals in Volume 1, section 9.6. If stationary, the statistical properties of the field (e.g. the average, higher order moments) do not depend on the origin of time. Similarly, for statistically homogeneous fields, their properties do not depend on the origin of space. In other words, correlations only depend on the difference between the space and time coordinates, and not on each coordinate specifically.

Wide-sense stationarity is less restrictive and defines a random process with only its first and second moments independent of the choice of origin. For the discussion here, the fields are assumed to be stationary at least in the wide sense. Under these circumstances, the dimensionality of the spatiotemporal correlation function Λ decreases by half, and we can write

$$\Lambda(\boldsymbol{\rho}, \tau) = \langle U(\mathbf{r}, t)U^*(\mathbf{r} + \boldsymbol{\rho}, t + \tau)\rangle_{\mathbf{r},t}, \tag{15.5}$$

where $\boldsymbol{\rho} = \mathbf{r}_2 - \mathbf{r}_1$ and $\tau = t_2 - t_1$.

Note that $\Lambda(\mathbf{0}, 0) = \langle U(\mathbf{r}, t)U^*(\mathbf{r}, t)\rangle_{\mathbf{r},t}$ represents the *spatially averaged irradiance* of the field, which is, of course, a real quantity. However, in general $\Lambda(\boldsymbol{\rho}, \tau)$ is complex. Let us define a normalized version of Λ, referred to as the *spatiotemporal complex degree of correlation*

$$\alpha(\boldsymbol{\rho}, \tau) = \frac{\Lambda(\boldsymbol{\rho}, \tau)}{\Lambda(\mathbf{0}, 0)}. \tag{15.6}$$

It can be shown that for stationary fields $|\Lambda|$ attains its maximum at $\mathbf{r} = 0$, $t = 0$, thus

$$0 \leq |\alpha(\rho, \tau)| \leq 1. \tag{15.7}$$

We formulated this result in section 10.2.1, when describing temporal noise. Note that, in homogeneous media, such as free space, the *dispersion relation* connects the three coordinates of the wave vector, $k^2 = k_x{}^2 + k_y{}^2 + k_z{}^2 = \omega^2/c^2$. As a consequence, in most cases, the *spatial* behavior of the optical field is well described in 2D, that is, in a plane. All measurements are performed either by 1D or 2D detectors; thus, we use the coherence *time* and *area* to characterize field statistics. We can define an area $A_C \propto \rho_C{}^2$ and length $l_c = c\tau_C$, over which $|\alpha(\rho_C, \tau_C)|$ maintains a significant value, say $|\alpha| > 1/2$, which defines a *coherence volume*

$$V_c = A_c l_c. \tag{15.8}$$

This *coherence volume* determines the maximum domain size over which the fields can be considered correlated. These parameters are of practical importance because they indicate over what spatiotemporal domain a field distribution maintains significant correlation with respect to a shifted replica of itself.

Figure 15.2 illustrates the measurement of temporal and spatial correlation functions using ensemble averaging. By comparison, for stationary fields, the procedure is significantly simplified, as shown in figure 15.3.

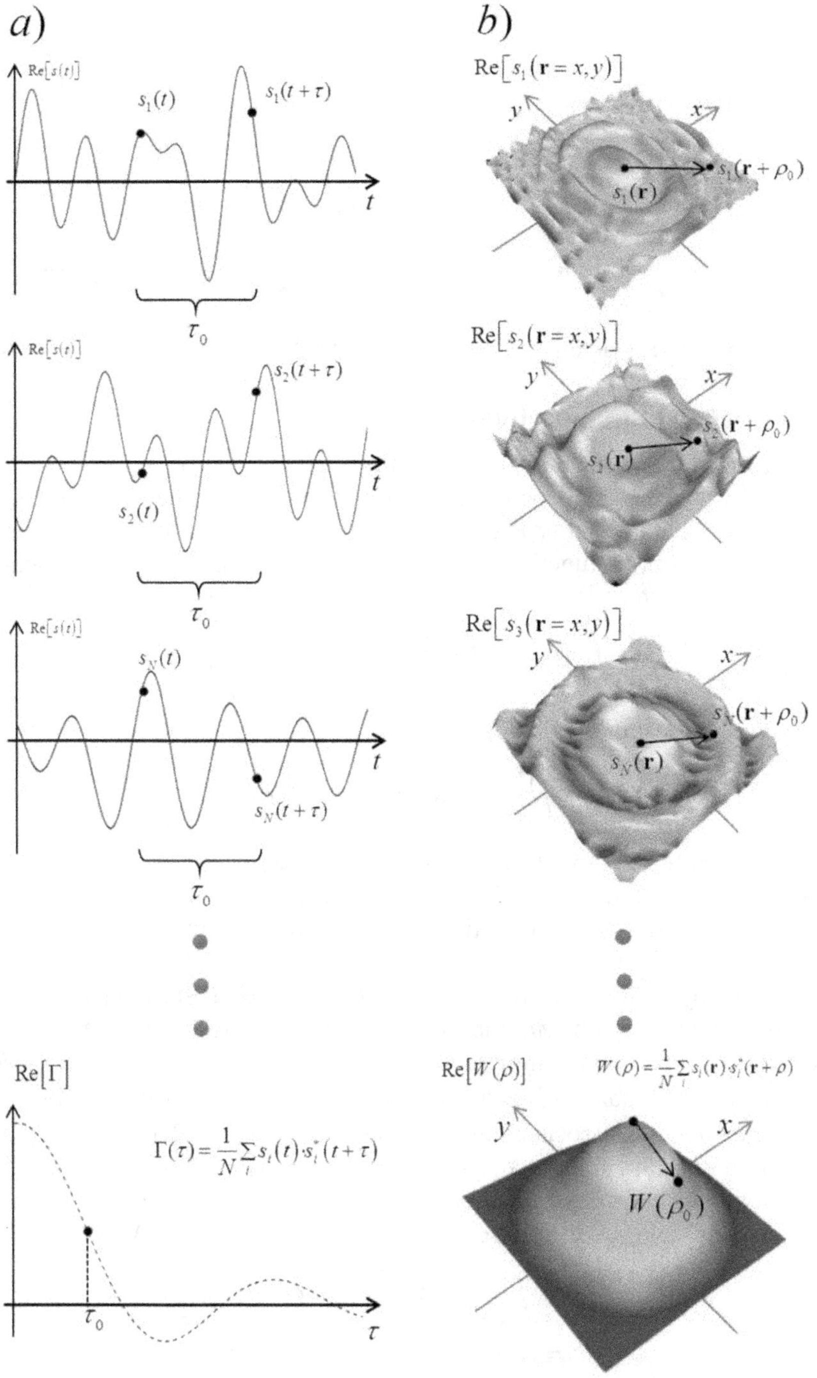

Figure 15.2. Illustrating the measurement of the temporal (a) and spatial (b) correlation function via ensemble averaging.

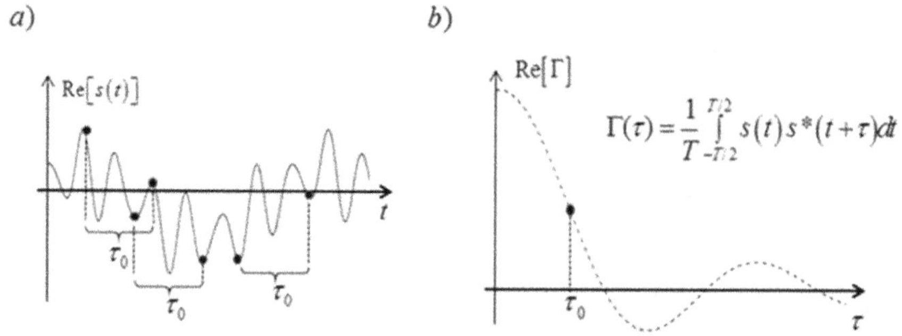

Figure 15.3. (a) Stationary field, $s(t)$. (b) The autocorrelation function of s measured as temporal averaging.

15.5 Wiener–Khintchine theorem

Generally, the random signal, $s(\mathbf{r}, t)$, does not have a Fourier transform in either time or space. However, independently, Wiener and Khintchine were able to prove mathematically that the *autocorrelation* of such signal does have a Fourier transform. Furthermore, this function is the *power spectrum* of the random signal. This relationship is known as the Wiener–Khintchine theorem (see chapter 10 in [5]),

$$\int_{-\infty}^{\infty}\int_V \Lambda(\boldsymbol{\rho}, \tau)\, e^{i(\omega\tau-\mathbf{k}\cdot\boldsymbol{\rho})} d^3\boldsymbol{\rho}\, d\tau = S(\mathbf{k}, \omega). \tag{15.9}$$

The inverse relationship also holds,

$$\Lambda(\boldsymbol{\rho}, \tau) = \int_{-\infty}^{\infty}\int_V S(\mathbf{k}, \omega)\, e^{-i(\omega\tau-\mathbf{k}\cdot\boldsymbol{\rho})} d^3\boldsymbol{\rho}\, d\tau. \tag{15.10}$$

In equations (15.9) and (15.10), we maintained our sign convention for the Fourier transform, introduced in Volume 1, section 3, whereby the space–time function (Λ in this case) is obtained as a superposition of monochromatic plane waves, denoted by $e^{-i(\omega\tau-\mathbf{k}\cdot\boldsymbol{\rho})}$.

Note that, by definition, the power spectrum of a stationary signal is a deterministic function with *real* and *positive* values. Because it is integrable, S can be normalized to the unit area to represent a *probability density*, $S(\mathbf{k}, \omega)/\int S(\mathbf{k}, \omega) d^3\mathbf{k}\, d\omega$. Its Fourier transform, a normalized version of Λ, is the *characteristic function* associated with a random process. Therefore, up to this normalization constant the autocorrelation function defined by equation (15.10) is nothing more but the *frequency-averaged* monochromatic plane wave associated with the random field, as seen in section 15.2,

$$\begin{aligned}\langle e^{-i(\omega t-\mathbf{k}\cdot\boldsymbol{\rho})}\rangle_{\mathbf{k},\omega} &\propto \int_{-\infty}^{\infty}\int_V S(\mathbf{k}, \omega)\cdot e^{-i(\omega\tau-\mathbf{k}\cdot\boldsymbol{\rho})} d^3\boldsymbol{\rho}\, d\tau \\ &= \Lambda(\boldsymbol{\rho}, \tau).\end{aligned} \tag{15.11}$$

Thus, the spatiotemporal correlation function has the very interesting physical interpretation of a *monochromatic plane wave*, averaged over all spatial and temporal frequencies.

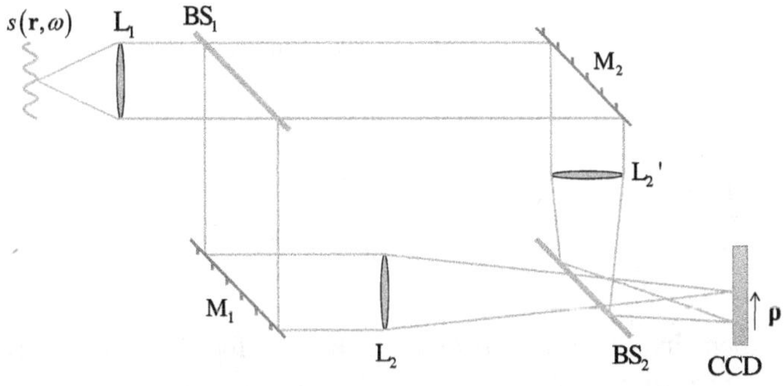

Figure 15.4. Mach–Zender interferometry with spatially extended fields for measuring $W(\boldsymbol{\rho})$: s is the source, $L_{1,2}$ the lenses, $M_{1,2}$ the mirrors, $BS_{1,2}$ the beam slitters, and CCD the charge-coupled device.

Note that for deterministic signals that have Fourier transforms, the Wiener–Khintchine theorem reduces to the *correlation theorem*, $f \otimes f \leftrightarrow |\tilde{f}|^2$. This is a general property of the Fourier transform, described in Volume 1, section 4.3. Therefore, the great importance of the Wiener–Khintchine theorem is due to its applicability to random signals, which do not possess a Fourier transform.

15.6 Spatial correlations of monochromatic light

15.6.1 Cross-spectral density

By taking the Fourier transform of equation (15.10) with respect to time, we obtain what is referred to as the (spatially-averaged) *cross-spectral density* [6]

$$\begin{aligned} W(\boldsymbol{\rho}, \omega) &= \int \Lambda(\boldsymbol{\rho}, \tau) \cdot e^{i\omega\tau}\, d\tau \\ &= \langle U(\mathbf{r}, \omega) \cdot U^* (\mathbf{r} + \boldsymbol{\rho}, \omega) \rangle_{\mathbf{r}}. \end{aligned} \tag{15.12}$$

The cross-spectral density function was used previously by Wolf to describe the *second-order statistics* of optical fields, that is, the Fourier transform of the temporal cross-correlation between two distinct points, $W_{12}(\mathbf{r}_1, \mathbf{r}_2, \omega) = \int \Gamma_{12}(\mathbf{r}_1, \mathbf{r}_2, \tau) \cdot e^{i\omega\tau}\, d\tau$ [1, 6]. This function describes the similarity in the field fluctuations of two points, such as, for example, in the two-slit Young interferometer. Note that two points are always fully correlated if the light is monochromatic, because, at most, the field at the two points can differ by a constant phase shift. However, across an entire plane, the phase distribution is a random variable. Therefore, in order to capture the spatial correlations in an *ensemble-averaged* sense, which is most relevant to imaging, we will use the spatially averaged version of $W(\boldsymbol{\rho}, \omega)$, as defined in equation (15.12).

One interferometric configuration that allows direct measurement of W is illustrated in figure 15.4 via an imaging Mach–Zehnder interferometer. Here the monochromatic field $U(\mathbf{r}, \omega)$ is split into two replicas that are further re-imaged at the CCD plane via two 4f lens systems, which induce a relative *spatial shift* $\boldsymbol{\rho}$. The question of practical interest here is: to what extent do we observe fringes, or, more quantitatively, what is the spatially averaged *fringe contrast* as we vary $\boldsymbol{\rho}$? For each value of $\boldsymbol{\rho}$, the CCD records a *spatially resolved* intensity distribution, or an *interferogram*. We can thus compute the *spatial average* of this quantity as

$$\begin{aligned} I(\boldsymbol{\rho}, \omega) &= \frac{1}{A}\int_A |U(\mathbf{r}, \omega) + U(\mathbf{r} + \boldsymbol{\rho}, \omega)|^2 \, d^2\mathbf{r} \\ &= \frac{1}{A}\int_A 2I(\mathbf{r}, \omega)d^2\mathbf{r} + 2\,\mathrm{Re}\frac{1}{A}\int [U(\mathbf{r}, \omega)\; U^*(\mathbf{r} + \boldsymbol{\rho}, \omega)]d_2\mathbf{r} \\ &= \langle I(\mathbf{r}, \omega)\rangle_{\mathrm{r}} + \mathrm{Re}\,[W(\boldsymbol{\rho}, \omega)], \end{aligned} \tag{15.13}$$

where we assumed that the interferometer splits the light equally on the two arms. Once the average irradiance of each beam, $\langle I(\mathbf{r}, \omega)\rangle_{\mathbf{r}}$, is measured separately (e.g. by blocking one beam and measuring the other), the real part of $W(\rho, \omega)$, as defined in equation (15.12), can be measured experimentally. Clearly, multiple CCD exposures are necessary to correspond to each $\boldsymbol{\rho}$. The *complex degree of spatial correlation* at frequency ω is defined as

$$\beta(\boldsymbol{\rho}, \omega) = \frac{W(\boldsymbol{\rho}, \omega)}{|W(\mathbf{0}, \omega)|}. \tag{15.14}$$

Note that $W(\mathbf{0}, \omega)$ is nothing more than the *average power spectrum* of the field,

$$\begin{aligned} W(\mathbf{0}, \omega) &= \langle U(\mathbf{r}, \omega) \cdot U^*(\mathbf{r}, \omega)\rangle_{\mathbf{r}} \\ &= \langle S(\mathbf{r}, \omega)\rangle_{\mathbf{r}}. \end{aligned} \tag{15.15}$$

Again, it can be shown that $|\beta| \in [0, 1]$, where the extremum values of $|\beta| = 0$ and $|\beta| = 1$ correspond to a complete lack of spatial correlation and full correlation, respectively. The area over which $|\beta|$ maintains a *significant value* defines the correlation area at frequency ω, for example

$$A_C = D, \quad \text{for which } |\beta(\boldsymbol{\rho}, \omega)|\Big| (\rho_x, \rho_y) \subset D < \frac{1}{2}. \tag{15.16}$$

Often, we refer to the *coherence area* of a certain field, without referring to a particular optical frequency. In this case, what is likely understood is the frequency-averaged correlation area, $A = \langle A_c(\omega)\rangle_\omega$. In practice, often we deal with fields that are characterized by a mean frequency, $\omega_0 = \int_{-\infty}^{\infty} \omega P(\omega)d\omega / \int_{-\infty}^{\infty} P(\omega)d\omega$. In this case the spatial coherence is well described by the behavior at this particular frequency. For example, a broadband field is fully spatially coherent if $|\beta(\boldsymbol{\rho}, \omega_0| = 1$, for any $\boldsymbol{\rho}$ in the domain [7, 8].

15.6.2 Spatial power spectrum

Since $W(\boldsymbol{\rho}, \omega)$ is a spatial correlation function, it can be expressed via a Fourier transform in terms of a *spatial power spectrum*, $P(\mathbf{k}, \omega)$, as described by the Wiener–Khintchine theorem (section 15.5),

$$\begin{aligned} P(\mathbf{k}_\perp, \omega) &= \iint W(\rho, \omega) \cdot e^{-i\mathbf{k}_\perp\rho}\, d^2\rho \\ W(\rho, \omega) &= \iint P(\mathbf{k}_\perp, \omega) \cdot e^{i\mathbf{k}\rho}\, d^2\mathbf{k}_\perp. \end{aligned} \tag{15.17}$$

It is straightforward to show that the spatial correlation of fields at two different frequencies vanishes,

$$\begin{aligned} W(\rho, \omega_1, \omega_2) &= \langle U_1(\mathbf{r}, \omega_1) \cdot U_2(\mathbf{r} + \rho, \omega_2)\rangle_\mathbf{r} \\ &= W(\rho, \omega_1)\delta(\omega_2 - \omega_1). \end{aligned} \tag{15.18}$$

The physical meaning of the term $\delta(\omega_2 - \omega_1)$ is that, performing the spatial correlation measurements in figure 15.4 with each field's different optical frequencies, ω_1, ω_2 (which may pass through two different filters), at each spatial shift $\boldsymbol{\rho}$, the spatial integration averages to zero, when the integration time at the detector is larger than $2\pi/(\omega_2 - \omega_1)$. Thus, maximum contrast fringes in a Mach–Zehnder interferometer such as that in figure 15.4 are obtained by having the same frequencies, $\omega_1 = \omega_2$, on both arms of the interferometer.

According to equation (15.17), the spatial correlation function $W(\boldsymbol{\rho}, \omega)$ can also be experimentally determined from measurements of the spatial power spectrum, as shown in figure 15.5. As we will discuss in more detail later, both the far field propagation in free space and propagation through a lens can generate the Fourier transform of the source field, as illustrated in figure 15.5,

$$\tilde{U}(\mathbf{k}_\perp, \omega) = \int_A U(\mathbf{r}, \omega) \cdot e^{-i\mathbf{k}_\perp\mathbf{r}}\, d^2\mathbf{r}. \tag{15.19}$$

The CCD is sensitive to power and, thus, detects the spatial power spectrum, $P(\mathbf{k}, \omega) = |\tilde{U}(\mathbf{k}, \omega)|^2$.

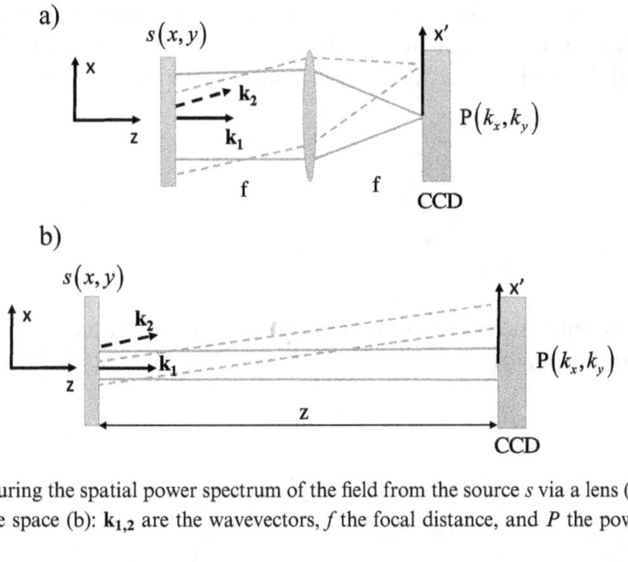

Figure 15.5. Measuring the spatial power spectrum of the field from the source s via a lens (a) and Fraunhofer propagation in free space (b): $\mathbf{k}_{1,2}$ are the wavevectors, f the focal distance, and P the power spectrum.

In equation (15.19), the frequency component $\mathbf{k} = (k_x, k_y)$ depends either on the focal distance, for the lens transformation (figure 15.5(a)), or on the propagation distance z, for the Fraunhofer propagation (figure 15.5(b)),

$$\begin{aligned} \mathbf{k}_{\perp}^{\text{lens}} &= \frac{2\pi}{\lambda f}(x', y') \\ \mathbf{k}_{\perp}^{\text{Fraunhofer}} &= \frac{2\pi}{\lambda z}(x', y'). \end{aligned} \tag{15.20}$$

Note that in the Fraunhofer regime, the ratios x/f and x/z describe the diffraction angle; therefore, sometimes $P(\mathbf{k}_{\perp}, \omega)$ is called the *angular power spectrum.*

An important question that arises equally often both in astronomy and microscopy is: how does the spatial correlation of the field change upon propagation? We will discuss propagation of field correlations much more rigorously later. For now, we seek to acquire an intuitive picture about this interesting phenomenon.

For extended sources that are far away from the detection plane, as in figure 15.5(b), the size of the source may have a significant effect on the Fourier transform in equation (15.19). This effect becomes obvious if we replace the source field U with its spatially truncated version, $\underline{U}$, to indicate the finite size of the source

$$\underline{U}(\mathbf{r}, \omega) = U(\mathbf{r}, \omega)\, \Pi\left(\frac{\mathbf{r}}{a}\right), \tag{15.21}$$

where the function Π is the typical 2D *rectangular function*, here denoting a square of side a. Thus, the far field becomes

$$\underline{\tilde{U}}(\mathbf{k}_{\perp}, \omega) = a^2 \tilde{U}(\mathbf{k}_{\perp}, \omega) \circledcirc_{\mathbf{k}_{\perp}} \operatorname{sinc}(a\mathbf{k}_{\perp}), \tag{15.22}$$

where $\circledcirc$ denotes convolution and *sinc* is the common $\sin(x)/x$ function (Volume 1, chapter 4). Thus, the field across the detection plane (x', y'), $\underline{\tilde{U}}(\mathbf{k}_{\perp}, \omega)$, is *smooth* over scales given by the width of the *sinc* function. This *smoothness* indicates that the field is spatially correlated over this spatial scale. Along x', this correlation distance, x_c', is obtained by writing explicitly the spatial frequency argument of the *sinc* function,

$$\begin{aligned} \frac{2\pi}{a} &= k_x \\ &= \frac{2\pi}{\lambda z} \cdot x_c'. \end{aligned} \tag{15.23}$$

We can conclude that the correlation area of the field generated by the source in the far zone is of the order of

$$\begin{aligned} A_C &= x_c^2 \\ &= \frac{\lambda^2}{\Delta\Omega}. \end{aligned} \tag{15.24}$$

In equation (15.24), Ω is the solid angle subtended by the source. This simple relationship allowed Michelson to measure interferometrically the angle subtended by stars. For example, the Sun subtends an angle $\theta \simeq 10$ mrad, that is, $\Omega \simeq 10^{-4}$ srad. Thus, for the green radiation that is the mean of the visible spectrum, $\lambda = 550$ nm, the coherence area at the surface of the Earth is of the order of $A_c^{\text{Sun}} = 50 \times 50\ \mu\text{m}^2$. Measuring this area over which the Sun's light shows correlations (or generates fringes) provides information about its angular size. For angularly smaller sources, far field spatial coherence is correspondingly higher. **This is the essence of the van Cittert–Zernike theorem, which states that the field generated by spatially incoherent sources gains coherence upon propagation.** This is the result of free-space propagation acting as a *spatial low-pass filter* [9].

It is perhaps not surprising that holding the knowledge of this powerful theorem, Zernike himself employed the spatial filtering concept to develop *phase contrast microscopy* [10, 11]. It had been known since the work of Abbe that an image can be described as an interference phenomenon [12]. Image formation is the result of simultaneous interference processes that take place at each point in the image. In a quest to make transparent specimens visible, Zernike employed spatial filtering and extended the coherence area of the illuminating field. Next, we will describe spatial filtering.

15.6.3 Spatial filtering

From the properties of Fourier transforms, we infer right away that higher spatial coherence at frequency ω, that is, a broader $W(\rho, \omega)$, can be obtained by narrowing $P(\mathbf{k}, \omega)$. When dealing with extended sources, it is common practice in the laboratory to perform low-pass filtering on $P(\mathbf{k}, \omega)$, such that the coherence area extends over the desired field of view. This procedure, commonly encountered in imaging experiments, is called, not surprisingly, *spatial filtering*, and is illustrated in figure 15.6.

In figure 15.6, the extended, source s emits light at a multitude of frequencies ω, and spatial frequencies $\mathbf{k}$. At a given frequency ω, lens L_1 performs the spatial Fourier transform. If an aperture is placed at this Fourier plane to block the high

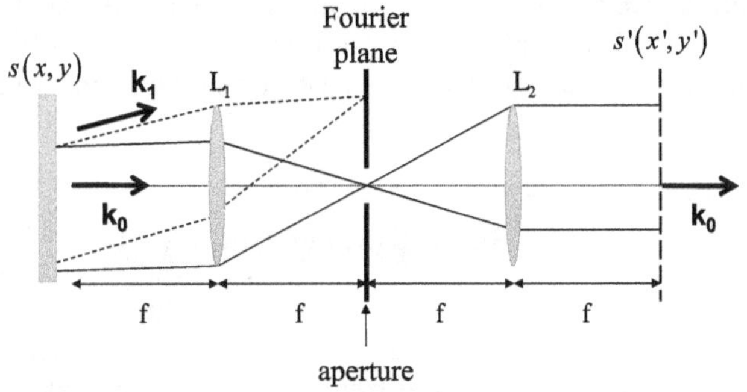

Figure 15.6. Spatial filtering via a 4f system. The field s' is a smoothed version of s, as the aperture performs a low-pass filtering operation.

spatial frequencies, the field reconstructed by lens L_2 at plane s' (conjugate to s) approximates a plane wave of wave vector $\mathbf{k}_0$. With this procedure, from an extended source, we obtain a highly spatial coherence field. Of course, this procedure is *lossy*, as the energy carried by the high spatial frequency is lost. It should be pointed out that all sources exhibit spatial coherence, at least at the scale of the wavelength. This is easily understood by noting that a δ-correlated source, $W(\rho) = W_0\delta(\rho)$, requires $\tilde{W}(\mathbf{k})$ to be infinitely broad, i.e. $\tilde{W}(\mathbf{k}) = 1$. Clearly, this is impossible, because a planar source can only emit in a 2π srad solid angle. Thus, the minimum coherence area for an arbitrary source is of the order of (equation (15.24))

$$A_C{}^{\min} \simeq \frac{\lambda^2}{2\pi}. \tag{15.25}$$

Physically, spatial coherence of a field over a given plane describes how close the field is to a plane wave. Alternatively, spatial coherence describes how well the field can be focused to a point (this point, of course, corresponds to a delta-function in the frequency domain). Spatial coherence plays an important role in microscopy and is used to differentiate between two classes of methods: (spatially) coherent versus incoherent.

15.7 Temporal correlations of plane waves

15.7.1 Temporal autocorrelation function

Let us now have the discussion analogue to that in section 15.6, where we now investigate the *temporal* correlations of fields at a particular spatial frequency $\mathbf{k}$ (or, equivalently, a plane wave propagating along a certain direction of propagation). Taking the *spatial* Fourier transform of Λ in equation (15.5) we obtain the *temporal correlation function*

$$\begin{aligned}\Gamma(\mathbf{k}, \tau) &= \iint \Lambda(\tilde{\rho}, \tau) e^{-i\mathbf{k}\rho}\, d^2\mathbf{r} \\ &= \langle U(\mathbf{k}, t) U^*(\mathbf{k}, t+\tau)\rangle_t.\end{aligned} \tag{15.26}$$

The autocorrelation function Γ is relevant in interferometric experiments of the type illustrated in figure (15.7). In a Michelson interferometer, a plane wave from the source is divided by the beam splitter and subsequently recombined via reflections on mirrors M_1, and M_2. The intensity at the detector has the form (we assume a 50/50 beam splitter)

$$\begin{aligned}I(\mathbf{k}, \tau) &= \langle |U(\mathbf{k}, t) + U^*(\mathbf{k}, t+\tau)|^2\rangle_t \\ &= 2I(\mathbf{k}) + 2\mathrm{Re}\langle U(\mathbf{k}, t)\, U^*(\mathbf{k}, t+\tau)\rangle.\end{aligned} \tag{15.27}$$

Thus the real part of $\Gamma(\mathbf{k}, \tau)$ is obtained by varying the time delay between the two fields. This delay can be controlled by translating one of the mirrors. The *complex degree of temporal correlation* at spatial frequency $\mathbf{k}$ is defined as

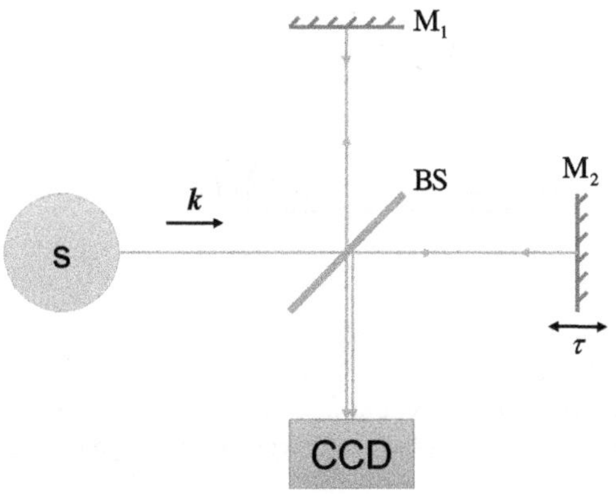

Figure 15.7. Michelson interferometry: s is the source, $\mathbf{k}$ the wavevector, $M_{1,2}$, the mirrors, BS the beam splitter, and τ the temporal delay introduced by displacing the mirror M_2.

$$\gamma(\mathbf{k}, \tau) = \frac{\Gamma(\mathbf{k}, \tau)}{|\Gamma(\mathbf{k}, 0)|}. \tag{15.28}$$

Note that $\Gamma(\mathbf{k}, 0)$ represents the intensity of the field, that is

$$\begin{aligned} \Gamma(\mathbf{k}, 0) &= \langle U(\mathbf{k}, t) U^* (\mathbf{k}, t) \rangle_t \\ &= I(\mathbf{k}). \end{aligned} \tag{15.29}$$

The complex degree of temporal correlation has a similar property with its *spatial* counterpart β, that is

$$0 \leq |\gamma(\mathbf{k}, \tau)| \leq 1. \tag{15.30}$$

Similarly, the *coherence time* is defined as the maximum time delay between the fields for which $|\gamma|$ maintains a significant value, say ½.

It is straightforward to show that if we cross-correlate temporally two plane waves of different wave vectors (directions of propagation), the result vanishes unless $\mathbf{k}_1 = \mathbf{k}_2$,

$$\begin{aligned} \Gamma(\mathbf{k}_1, \mathbf{k}_2, \tau) &= \langle U_1(\mathbf{k}_1, t) U_2^* (\mathbf{k}_2, t + \tau) \rangle_t \\ &= \Gamma(\mathbf{k}_1, \mathbf{k}_1, \tau) \delta(\mathbf{k}_2 - \mathbf{k}_1). \end{aligned} \tag{15.31}$$

Thus, at each moment t, the two plane waves generate fringes perpendicular to $\mathbf{k}_2 - \mathbf{k}_1$, which average the signal to zero. Similar to the spatial domain case, when the spatial correlation at two different temporal frequencies was discussed (section 15.6.1), if the detector (e.g. a CCD) averages the signal over scales larger than the fringe period, the temporal correlation information is lost. As τ changes, the fringes 'run' across the plane such that the contrast averages to 0. For this reason, the two beams in a typical Michelson interferometer are carefully aligned to be parallel.

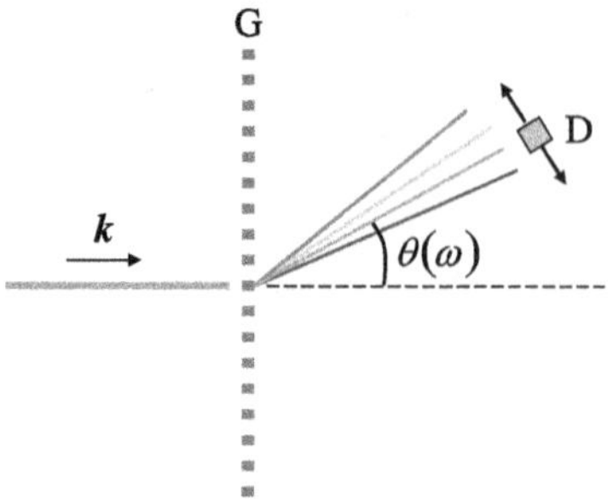

Figure 15.8. Spectroscopic measurement using a grating: G is the grating, D is the detector, and θ the diffraction angle. The dashed line indicates the undiffracted order (zeroth order).

15.7.2 Optical power spectrum

The temporal correlation Γ is the Fourier transform of the power spectrum,

$$\begin{aligned} \Gamma(\mathbf{k}, \tau) &= \int_{-\infty}^{\infty} S(\mathbf{k}, \omega)\, e^{-i\omega\tau}\, d\omega \\ S(\mathbf{k}, \omega) &= \int_{-\infty}^{\infty} \Gamma(\mathbf{k}, \tau)\, e^{i\omega\tau}\, d\omega. \end{aligned} \tag{15.32}$$

Thus, Γ can be determined via spectroscopic measurements, as exemplified in figure 15.8. By using a grating (a prism, or any other dispersive element), we can 'disperse' different colors at different angles, such that a rotating detector can measure $S(\omega)$ directly. In order to estimate the coherence time for a broadband field, let us assume a Gaussian spectrum centered at frequency ω_0, and having the standard deviation width $\Delta\omega$,

$$S(\omega) = S_0 e^{-\left(\frac{\omega-\omega_0}{\sqrt{2}\Delta\omega}\right)^2}, \tag{15.33}$$

where S_0 is a constant. Using the Fourier transform properties of a Gaussian function (Volume 1, chapter 4), the autocorrelation function is also a Gaussian, modulated by a sinusoidal function, as a result of the Fourier shift theorem

$$\Gamma(\tau) = \Gamma_0 e^{-\left(\frac{\Delta\omega\tau}{\sqrt{2}}\right)^2} e^{i\omega_0\tau}. \tag{15.34}$$

From equation (15.34), we see that if we define the width of Γ as the coherence time, we obtain

$$\tau_c \propto \frac{1}{\Delta\omega}, \tag{15.35}$$

and the coherence length

$$\begin{aligned} l_c &= c\tau_C \\ &\propto \frac{\lambda^2}{\Delta\lambda}. \end{aligned} \tag{15.36}$$

The coherence length depends on the spectral bandwidth in an analog fashion to the coherence area dependence on the solid angle (equation (15.24)). This is not surprising as both types of correlations depend on their respective frequency bandwidth.

15.7.3 Spectral filtering

The coherence length can vary broadly, from kilometers for a narrow band laser, to microns for LEDs and white light. Figure 15.9 shows qualitatively the relationship between $\Gamma(\tau)$ and $S(\omega)$. The short coherence length of a broad band source is the starting point in *low-coherence interferometry* and *optical coherence tomography* [13], as discussed later. Of course, using narrow band filters has the effect of enlarging the coherence length of the field.

15.8 Spatially-dependent coherence time and temporally-dependent coherence area

Let us consider the spatiotemporal fluctuations of a field observed at a given plane (x, y). This is a common situation encountered in practice where, for example, a field emitted by a primary source is investigated at a plane, perpendicular to the z-axis (figure 15.10). We define the spatial and temporal bandwidths as standard deviations of the spatiotemporal power spectrum, σ_k, σ_ω assumed to be normalized to the probability density, namely

$$\int_{-\infty}^{\infty}\int_{A_{\mathbf{k}_\perp}} S(k_\perp, \omega)d^2\mathbf{k}_\perp\, d\omega = 1. \tag{15.37}$$

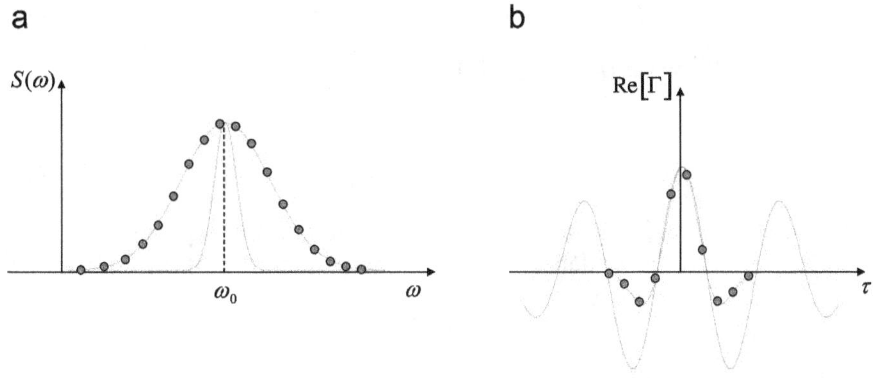

Figure 15.9. (a) Broad (dashed line) and narrow (solid line) power spectrum. (b) Temporal autocorrelation functions associated with the power spectra in (a).

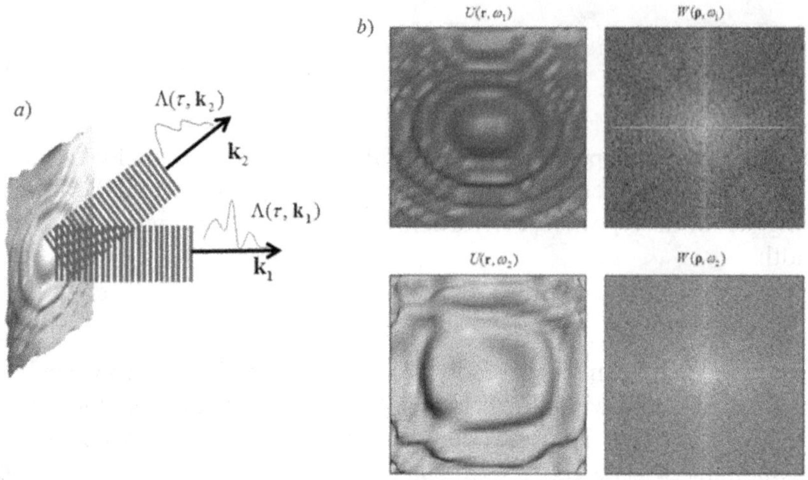

Figure 15.10. (a) Different temporal correlation functions for two plane waves of wavevectors $\mathbf{k_1}$ and $\mathbf{k_2}$. (b) Different cross-spectral densities for two optical frequencies, ω_1 and ω_2.

In equation (15.37), $A_{k_\perp}$ is the domain of all possible values for the transverse wave vector, $\mathbf{k}_\perp = (k_x, k_y)$. We define the coherence time, τ_c, and coherence areas, A_c, as the inverse of their respective power spectra,

$$\tau_c = \frac{1}{\sigma_\omega} \tag{15.38}$$

$$A_c = \frac{1}{\sigma_{k_\perp}^2}. \tag{15.39}$$

For simplicity, we assume that the spatial statistics is isotropic, in particular, $\sigma_{k_\perp}^2 = \sigma_{k_x}^2 + \sigma_{k_y}^2$. The two variances are defined as

$$\begin{aligned}\sigma_\omega^2(\boldsymbol{k}_\perp) &= \langle(\omega - \langle\omega\rangle)^2\rangle \\ &= \int_{-\infty}^{\infty} \omega^2 S(\boldsymbol{k}_\perp, \omega)d\omega - [\int_{-\infty}^{\infty} \omega S(\boldsymbol{k}_\perp, \omega)d\omega]^2 \\ &= \langle\omega^2(\boldsymbol{k}_\perp)\rangle - \langle\omega(\boldsymbol{k}_\perp)\rangle^2\end{aligned} \tag{15.40}$$

and

$$\sigma_{k_\perp}^2(\omega) = \langle|(k_\perp - \langle k_\perp\rangle|^2\rangle. \tag{15.41}$$

Clearly, the temporal bandwidth, σ_ω, depends on the spatial frequency $\boldsymbol{k}_\perp$. The physical meaning of a $\boldsymbol{k}_\perp$-dependent coherence time is that each plane wave component of the field can have a different temporal bandwidth and, thus, coherence time, $\tau_c(\boldsymbol{k}_\perp) = \frac{1}{\sigma_\omega(\boldsymbol{k}_\perp)}$, as illustrated in figure 15.10(a). Conversely, each monochromatic component can have a different spatial bandwidth and, thus, *coherence areas,* $A_c(\omega) = \frac{1}{\sigma_{k_\perp}^2(\omega)}$, as shown in figure 15.10(b).

The two variances can be further averaged with respect to these respective variables, such that they become constant.

$$\langle\sigma_\omega{}^2\rangle_{k_\perp} = \int_{A_{k_\perp}} \sigma_\omega{}^2(\boldsymbol{k}_\perp) S(\boldsymbol{k}_\perp, \omega) d^2 k_\perp \tag{15.42}$$

$$\langle\sigma_{k_\perp}{}^2\rangle_\omega = \int_{-\infty}^{\infty} \sigma_{k_\perp}{}^2(\omega) S(k_\perp, \omega) d\omega. \tag{15.43}$$

Thus, equation (15.42) yields a coherence time, $\tau_c = \frac{1}{\langle\sigma_\omega{}^2\rangle_{\mathbf{k}_\perp}}$, which is averaged over all spatial frequencies, while equation (15.43) provides a coherence area, $A_c = \frac{1}{\sigma_{k_\perp}{}^2}$, which is averaged over all temporal frequencies.

Interestingly, although we always deal with fields that fluctuate in both time and space, rarely do we specify a τ_c as a function of $\boldsymbol{k}_\perp$ or A_c versus ω. This happens because we implicitly assume averaging of the form in equations (15.42) and (15.43). In many situations, the power spectrum is narrow. Thus, we can use the *quasi-monochromatic approximation,* which states that the temporal bandwidth is much smaller than the mean frequency, $\frac{\sigma_\omega}{\langle\omega\rangle} \ll 1$. Conversely, we can define a *quasi-phase wave* as a field characterized by a spatial bandwidth that is much smaller than the mean wave vector magnitude, $k = \sqrt{k_x{}^2 + k_y{}^2 + k_z{}^2}$, namely, $\sigma_{k_\perp} \ll k$.

In the particular case when the optical spectrum is the same at each point in space, or when the spatial spectrum is constant in time, the spatiotemporal power spectrum factorizes as

$$S(\boldsymbol{k}_\perp, \omega) = S_{k_\perp}(\boldsymbol{k}_\perp) S_\omega(\omega). \tag{15.44}$$

It is easy to see that, in this case, the respective bandwidths are constant, namely,

$$\sigma_\omega^2(k_\perp) = \langle\sigma_\omega^2(k_\perp)\rangle_{k_\perp} \tag{15.45}$$

$$\sigma_{k_\perp}^2(\omega) = \langle\sigma_{k_\perp}^2(\omega)\rangle_\omega. \tag{15.46}$$

In summary, equations (15.42)–(15.43) provide a procedure for calculating the coherence time and areas for the general case of an optical field that fluctuates randomly in both time and space.

15.9 Problems

1. An incandescent filament of length $l = 10$ mm and width $w = 1$ mm emits light of wavelength $\lambda = 700$ nm (figure 15.11). Describe the spatial coherence of the radiated field at a screen placed $L = 1$ m, 10 m, and 100 m away.
2. How do the results in problem 1 change if the filament is covered by an aperture of diameter $d = 0.5$ mm, 1 mm, 5 mm (figure 15.12) ?
3. Calculate the coherence length of a field with a uniform power spectrum of central wavelength $\lambda_0 = 1$ μm and width $\Delta\lambda = 10$ nm, using the following conventions:

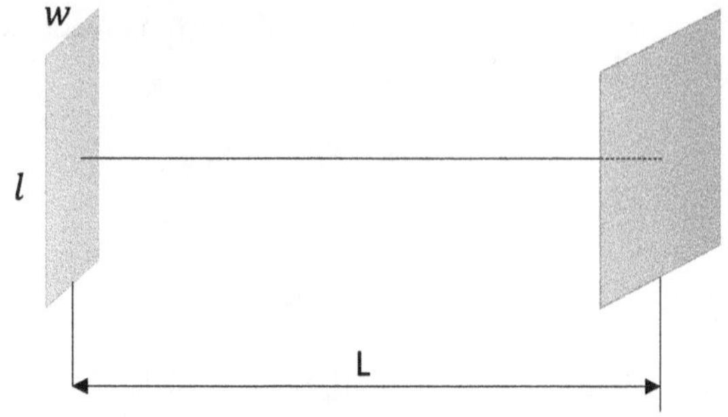

Figure 15.11. Problem 15.1.

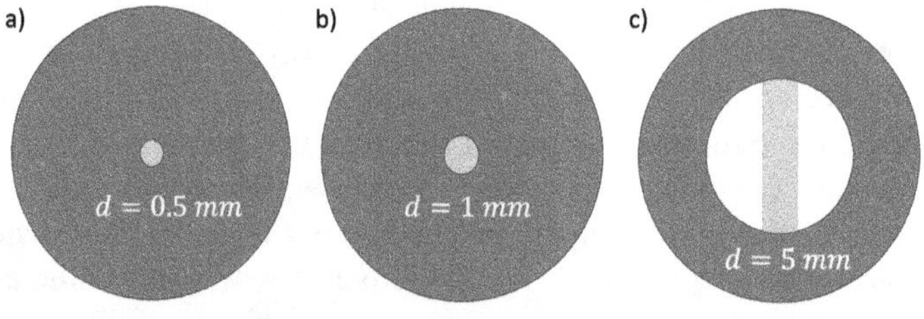

Figure 15.12. Problem 15.2: (a) d = 0.5 mm, (b) d = 1 mm, (c) d = 5 mm. The width of the filament is w = 1 mm, and its length is L = 10 mm.

 a) assume the coherence time as full-width half maximum of the autocorrelation function;
 b) assume the coherence time as the first root of the autocorrelation function;
 c) assume the coherence time as the inverse standard deviation of the frequency bandwidth.

4. A halogen lamp emits white light, which is well approximated by a Gaussian spectrum of central wavelength $\lambda_0 = 550$ nm and standard deviation $\Delta\lambda = 200$ nm.
   a) Calculate the coherence length of the white light field.
   b) A color filter passes light with a Gaussian spectrum, now with a central wavelength $\lambda_1 = 600$ nm and a bandwidth $\Delta\lambda_1 = 10$ nm. Calculate the new coherence length.
   c) Same as (b), but now the filter passes light with $\lambda_2 = 400$ nm, $\Delta\lambda_2 = 10$ nm.
   d) The light from (b) and (c) is combined. What is the coherence length for the composite light?

References

[1] Mandel L and Wolf E 1995 *Optical Coherence and Quantum Optics* (Cambridge: Cambridge University Press), xxvi p 1166

[2] Goodman J W 2000 *Statistical Optics* (New York: Wiley), xvii p 550

[3] Glauber R J 1963 The quantum theory of optical coherence *Phys. Rev.* **130** 2529

[4] Papoulis A 1991 Probability, random variables, and stochastic processes 3rd edn (McGraw-Hill Series in Electrical Engineering Communications and Signal Processing) (New York: McGraw-Hill), xvii p 666

[5] Born M and Wolf E 1999 *Principles of Optics: Electromagnetic Theory of Propagation, Interference and Diffraction of Light* 7th edn (Cambridge: Cambridge University Press), xxxiii p 952

[6] Wolf E 1982 New theory of partial coherence in the space-frequency domain. 1. Spectra and cross spectra of steady-state sources *J. Opt. Soc. Am.* **72** 343–51

[7] Wolf E 2009 Solution of thc phasc problem in the theory of structure determination of crystals from x-ray diffraction experiments *Phys. Rev. Lett.* **103** 075501

[8] Mandel L and Wolf E 1981 Complete coherence in the space-frequency domain *Opt. Commun.* **36** 247–9

[9] Goodman J W 1996 Introduction to fourier optics 2nd edn (McGraw-Hill Series in Electrical and Computer Engineering) (New York: McGraw-Hill), xviii p 441

[10] Zernike F 1942 Phase contrast, a new method for the microscopic observation of transparent objects, part 1 *Physica* **9** 686–98

[11] Zernike F 1942 Phase contrast, a new method for the microscopic observation of transparent objects, part 2 *Physica* **9** 974–86

[12] Abbe E 1873 Beiträge zur Theorie des Mikroskops und der mikroskopischen Wahrnehmung *Arch. Mikrosk. Anat.* **9** 431

[13] Huang D *et al* 1991 Optical coherence tomography *Science* **254** 1178–81

www.ingramcontent.com/pod-product-compliance
Lightning Source LLC
Chambersburg PA
CBHW080543220326
41599CB00032B/6342
* 9 7 8 0 7 5 0 3 1 9 5 1 5 *